Physiological Determinants
of Crop Growth

Physiological Determinants of Crop Growth

D. A. Charles-Edwards

CSIRO Division of Tropical Crops and Pastures
Cunningham Laboratory, St. Lucia, Queensland

ACADEMIC PRESS

A Subsidiary of Harcourt Brace Jovanovich, Publishers

Sydney   New York   London
Paris   San Diego   San Francisco   São Paulo   Tokyo   Toronto

ACADEMIC PRESS AUSTRALIA
Centrecourt, 25–27 Paul Street North
North Ryde, N.S.W. 2113

United States Edition published by
ACADEMIC PRESS INC.
111 Fifth Avenue
New York, New York 10003

United Kingdom Edition published by
ACADEMIC PRESS, INC. (LONDON) LTD.
24/28 Oval Road, London NW1 7DX

Printed in Australia

National Library of Australia Cataloguing-in-Publication Data

Charles-Edwards, D. A., 1944-.
  Physiological determinants of crop growth.

  Bibliography.
  Includes index.
  ISBN 0 12 169360 0.

  1. Plant physiology. 2. Growth (Plants).
  3. Plants, Cultivated. I. Title.

581.1

Library of Congress Catalog Card Number: 82-72698

For Bethan, Catrin and Elin

O Lord, open our eyes
to see what is beautiful;
Or minds to know what is true;
Our hearts to love what is good.

Contents

4
Light-Utilization Efficiency

5
Dry-Matter Partitioning

6
Implications, Limitations and Applications

Appendixes

Preface

Knowledge about something is not necessarily synonymous with an understanding of it. By knowledge I mean the cognizance of a proven fact or observation, and by understanding I mean explanation based on knowledge. Thus, although we may have a considerable body of knowledge about crops and agricultural systems, often our understanding of them is very limited. Throughout recorded history, man has applied his knowledge successfully to improve both the biological and economic efficiency of his agricultural systems. However, there is now increasing evidence that the returns on the application of new and existing knowledge are diminishing: the productivity of many agricultural systems seems to be near its maximum. Therefore, it becomes increasingly important that we develop and exploit our understanding of these cropping systems. For example, we may know that one cultivar of soybean produces more than another under a particular management regimen. If we understand why it outperforms the other cultivar, we might be able to improve the productivity of both cultivars through genetic or managerial manipulations of them.

In recent years there has been a growing awareness of the need to establish some kind of framework within which we can order our knowledge and thereby develop a proper understanding of the factors affecting the productivity of agricultural systems. I hope that this book illustrates one way in which mathematics can be used for this purpose. The analysis developed in it identifies five determinants of both the rate and extent of growth: the efficiency with which the crop uses intercepted light energy in the production of new dry matter; the amount of light energy intercepted by the crop; the partition of new crop dry matter between the different plant parts; the loss of dry matter through unavoidable physiological causes; and the duration of production of dry matter. It extends the type of analysis originally outlined by John Cooper, John Warren Wilson and John Monteith so that it can be more effectively used in the routine analysis of crop growth data.

The philosophy of the approach, and the constraints upon it, are outlined in the first chapter. In the second chapter, a variety of published data are used to illustrate the practical and theoretical bases of the analysis. These data are for both field crops and single plants. Three of the determinants of crop growth are then examined in some detail, and a more fundamental physiological analysis of them is developed. In Chapter 3 the interception of incident light energy by spaced plants, "closed" crops and rows of plants is examined. Simple relationships describing seasonal changes in the incident light and temperatures of sites at different latitudes are also proposed. Chapter 4 is concerned with the light-utilization efficiency. Its dependence on the photosynthetic capacity of the crop, the crop's respiratory activity and the canopy architecture is examined. The factors affecting the partition of new dry matter between different plant parts are examined in Chapter 5. Some of the problems of extending this sort of analysis to extensive cropping systems, such as pastures, and in defining a suitable minimum experimental data core are examined in Chapter 6.

The mathematical analyses described in this book are not definitive; they represent no more than stepping-stones on the way toward building a framework. I have no doubt that they will change with usage and with experience. For the most part, the mathematical manipulations used are no more than straightforward algebra and elementary calculus. The ideas that the mathematical equations attempt to formalize are not new, but several different ideas are brought together and used to build one framework within which we can develop an understanding of the physiological determinants of the growth of plants and crops.

In general I have used data with which I am familiar to illustrate my arguments. There are numerous sets of published data that I could have used; indeed a casual computer-aided search of the literature published during the past ten years identified more than five hundred papers relevant to these analyses. I make no apology for restricting myself to the data I know the best. In 1820, the chemist Samuel Parkes wrote, in his "Chemical Catechism", the following commendation for his science: "A knowledge of the chemical properties of bodies will thus give a new character to the agriculturalist, and render his employment rational and respectable". If the reader feels that my style of writing seems arrogant or pompous, I will plead that I am following in a long-established chemical tradition.

I have thoroughly enjoyed putting these ideas together, and I hope that colleagues find equal enjoyment in reading the book and putting the ideas contained in it to the test.

Acknowledgements

I am indebted to colleagues in this Division for their patience with me, and for the help that they have afforded me. Ted Henzell, Ian Wood, Bob McCown, Bob Myers, Bob Lawn and Peter Ross each read all or part of the manuscript for me and made time to debate it with me. Although I have not heeded all of their advice, neither have I ignored it. I thank them all for their hard work and diligence. The responsibility for the written word rests fairly upon my shoulders. I must record my special thanks to Ted Henzell, Chief of the Division, who gave me the opportunity to develop the ideas contained herein and to assemble them. I hope that this book fulfils his hopes and expectations.

I must thank Anne Ting for coping with my handwriting and having considerable patience in dealing with my innumerable redrafts, additions and deletions when preparing the manuscript for me.

Ken Rousell prepared the art-work for me, and Dallas Cox of Academic Press Australia edited the manuscript. I thank them both for their hard work.

Finally, I wish to thank the publishers of the following publications for permission to reproduce some of the illustrations used in this book:

Fig. 3.10 — *Annals of Applied Biology*, 1978, **90**.

Fig. 3.11 — *Annals of Botany*, 1976, **40**.

Fig. 4.7 — *Annals of Botany*, 1980, **46**.

Figs. 5.6 and 5.7 — *Australian Journal of Agricultural Research*, 1982, **33**.

1

Introduction

1.1 Objectives of an analysis

Synopsis: The acquisition of knowledge of the potential yield of a crop and the understanding of factors affecting that yield are objectives central to almost all agricultural research programs. Statistics has traditionally played an important role in helping to attain them. Mathematics has another, complementary role. Mathematical models allow us to formalize hypotheses about crop performance and about environmental effectors of crop performance. Although dynamic, mathematical simulation models can be resource-demanding and often have limited practical value, analytical mathematical models may provide a simple and direct approach in elucidating the effectors of potential crop yield. The analyses are subject to practical constraints, but they may provide a useful, complementary tool to the traditional methods of crop assessment.

A knowledge of the potential yield of a crop and the practical limitations to its realization in the field are essential prerequisites before any decision can be made on the crop's economic viability (Cooper, 1970). The acquisition of this knowledge about particular crops is central to almost all research programs in the agricultural sciences. We need to be able to identify those morphological and physiological characters which are most closely associated with high crop yields, and we need to know how they are affected by

1

both environmental and management changes in the cropping regimen. Traditionally, these needs have been met by growing each crop that is of interest under a range of contrasting conditions, and then comparing its performance under each growth regimen by direct observation. The statistical techniques of correlation analysis have usually been used to help establish which of the identifiable environmental, morphological or physiological characters are most highly correlated with any observed differences in crop yield. Whilst the approach is reasonably safe and has the important virtue of simplicity, it is demanding on both time and other experimental resources. As we learn more about the physiological and environmental effectors of plant growth, it becomes attractive to supplement this traditional, pragmatic approach with a complementary, deductive one.

The traditional approach has been essentially descriptive and has rightly given a great impetus to the development of the descriptive mathematical techniques of statistics. The rapid development of statistical theory and methodology in the first half of this century provided inestimable help to the agricultural sciences. The great variety of crop plants, cropping environments and cropping strategies that existed needed to be described and then related, one to the other. Until this was done and some semblance of order was created from the large body of experimental information, it was difficult to rationally describe, let alone understand, the underlying relationships between crop performance and the cropping environment. Once relationships had been indicated, statistics were then used to test, both qualitatively and quantitatively, hypotheses about them. The needs of the agricultural sciences have been, and still are, partially met by the statistical techniques of multifactorial experimentation and multiple regression analysis. However, statistical techniques do not distinguish between causal and casual relationships, and we need some framework within which we can order the information we have obtained. Then we need to use our understanding of plant growth processes to distinguish the causal relationships from the casual ones.

Mathematics can be used in another way. Whereas statistics is essentially descriptive mathematics, mathematics can also be used as a tool to analyse crop growth into its constituent processes, or to synthesize a knowledge of those processes into a predictor of crop growth. In the agricultural sciences the techniques of mathematical modelling are often seen as a discipline apart from, and different to, the mainstream of research activities, a view that is too often reinforced by the impressive and apparently erudite jargon that is associated with modelling activities. In a challenging article entitled "Sense and nonsense in crop simulation", Passioura (1973) examined some of the problems associated with the development of dynamic, mathematical crop simulation models. He attributed these problems to the difficulties encountered when we attempt to integrate knowledge up through the different levels of biological organization of a crop. Implicit in his conclusions was the idea that crop simulation models represented an activity that was different to, and separate from, the mainstream of agricultural research. This apparent dichotomy between crop simulation and agricultural research was clearly stated in his concluding paragraph:

> If research is the art of the soluble, computer simulation is the art of the plausible. As such it is closer to metaphysics than it is to science.

It is worthwhile to pause here and examine the roles that mathematics and mathematical models can play in helping us to understand the systems that we are researching.

A model, any model, whether it is physical or abstract, is no more than a simplified representation of the reality it purports to describe. We are familiar with the use of physical models, of objects such as ships or aeroplanes, to help us understand how the real object might behave in specific situations. The uses of abstract mathematical models of real systems are no different from these uses of physical models. Mathematics simply provides a language with which we can articulate and formalize our ideas about the system in an abstract way. We can use mathematics in essentially two ways. Firstly, we can use it to put together what we believe are the basic components of the system (system synthesis), and then we can see whether or not, when assembled, our mathematical model describes the behaviour of the real system. By analogy with physical models, mathematics is the glue that we use to stick the parts of the model, that is our assumptions about the system, together. If our model of the system does describe the behaviour of the real system, then we can use it to predict how the real system might behave under different circumstances. Usually this type of model is concerned with predicting how the system changes with time and, in the jargon of the modeller, it is a dynamic, mathematical simulation model of the real system. Secondly, we can use our mathematics to analyse the behaviour of the real system, and to help us deduce and identify its main components. These two uses of mathematics are not mutually exclusive; they are complementary. This book is concerned with the second type of analytical, mathematical model, but the analyses that are described here may also provide a framework within which we could build a dynamic, simulation model of a particular agricultural system.

Whilst dynamic simulation models have considerable social and economic potential if their ability to accurately predict crop response to environment and management practice is proven, their most immediate value to the research scientist are as heuristic aids to the understanding of the performance of the systems that they attempt to describe; that is, they may suggest the type of behaviour that might be expected from the system. Since a mathematical model is only a formalized statement of the assumptions that have been made about the system it describes, it can help us to identify those areas where our understanding of the system (our assumptions about it) and our knowledge are inadequate. It is a matter of common sense that, if our mathematical model makes a prediction of crop production based on an incomplete knowledge of basic crop physiological processes, it will need considerable experimental, scientific support at that basic physiological level. It will need to integrate our fundamental physiological knowledge, and our understanding of plant processes derived from that knowledge, across several different levels of plant and community organization, and we will need to independently check that it behaves correctly at each of these intermediate levels of organization. Alternatively, we can use analytical mathematical models to analyse crop production directly in terms of plant behaviour at these intermediate levels, and then to develop further, more detailed models to analyse the behaviour at these levels in terms of the more basic physiological processes. If we choose the latter approach, our field, agricultural experiments will begin to seek answers to the questions "how?" and "why?", the objectives of our research program will become more clearly defined, and they will be directed toward a more basic understanding of the main determinants of crop growth and production. In this new situation, simple mathematics provides the experimental agriculturist with a powerful and necessary tool to help him attain his objectives.

There are two main requirements of this type of analytical mathematical approach. Firstly, it needs to have simple and direct applications to the agricultural research problems at hand and, secondly, it needs to be able to be extended so that it can be used to integrate our understanding of the fundamental physiological properties of plants into a proper understanding of their performance at the agricultural field level.

1.2   A logical framework for the analysis of crop growth data

Synopsis: Using simple, logical analysis we can write the *net* daily rate of crop dry-matter production, $\Delta W/\Delta t$, as

$$\Delta W/\Delta t = \varepsilon J - V$$

where J is the amount of light energy intercepted by the crop during the day, ε is the efficiency with which the crop uses that light energy in the production of *new* dry matter, and V is the daily rate of loss of crop dry matter. The yield of the harvestable component, W_H, can be written as

$$W_H = \sum^{i=t} (\eta_H \varepsilon J - V_H)_i$$

where η_H is the proportion of the daily increment in *new* dry matter partitioned to the harvestable component, V_H is its daily rate of dry-matter loss and t is the duration of the period of its production. The analyses have important implications to the experimental study of both environmental and physiological effectors of the rate of crop growth and the potential yield of the crop.

Warren Wilson (1971) has outlined a physiological basis for the analysis of the maximum yield potential of arable crops in which he has proposed that the *net* growth rate of a "closed" crop, $\Delta W/\Delta t$, can be written as

$$\Delta W/\Delta t = \varepsilon^* \bar{I}[1 - \exp(-kL)] \qquad (1.1)$$

where ε^* is the efficiency with which incident light energy is used by the crop in the *net* production of dry matter, $\bar{I}$ is the daily average of the downward light flux density incident upon the crop's uppermost surface, k is a canopy light extinction coefficient and L is the crop's leaf area index. A "closed" crop can be defined as one in which the spatial heterogeneity in the downward light flux density in any horizontal plane within the volume of the crop canopy can be ignored. For a "closed" crop the term $[1 - \exp(-kL)]$ describes the proportion of the downward light flux density incident upon its uppermost surface that is absorbed by it. The term "light" can be defined as pertaining to that portion of the incident radiant energy that lies within the 400–700 nm waveband, the conventional definition of photosynthetically active radiation. This definition of the word "light" is used throughout the book.

Now, losses of dry matter, through physiological, pathological and mechanical (lodging and grazing) causes, are implicit in the parameter ε^*. We can generalize eqn (1.1), and explicitly take these losses into account, by re-writing the *net* growth rate of a crop as

$$\Delta W/\Delta t = \varepsilon J - V \qquad (1.2)$$

where ε is the efficiency with which *new* dry matter is produced, J is the amount of light energy absorbed by the crop during the increment of time Δt, and V is the rate of loss of dry matter by all causes—physiological, pathological (pests and diseases) and mechanical (lodging and grazing). The *net* amount of dry matter produced over the time interval t days, $W(t)$, can then be obtained from eqn (1.2) by summing all increments in dry matter,

ΔW, over time, which gives us

$$W(t) = \sum_{}^{i=t} (\varepsilon J - V)_i + W(0) \tag{1.3}$$

where $W(0)$ is the initial amount of dry matter present at time $t = 0$. If we now define the maximum yield potential of a crop as the maximum harvestable yield of its economic component(s) under optimal cropping regimens, when physiological, pathological or mechanical losses of dry matter can be minimized, we can use eqn (1.3) to write the maximum yield potential of a crop, W_H, as

$$W_H = \sum_{}^{i=t} (\eta_H \varepsilon J - V_H)_i \tag{1.4}$$

where η_H is the proportion of the daily increment in new crop dry matter partitioned to the harvestable parts, V_H is the rate of loss of dry matter by those parts through all unavoidable causes, and t becomes the duration of production of the harvestable part of the crop. Equation (1.4) simply states that the total yield of the harvestable component is the sum of all the *net* daily increments in dry matter of that component during the period of its production. Equation (1.4) defines five distinct physiological determinants of the maximum yield potential of a crop; *viz.* ε, J, η_H, V_H and t. These determinants are identified by simple logic. It needs to be emphasized that they are defined at the crop/plant level of organization. They are not defined in terms of physiological processes occurring at some lower level of biological organization, but in terms of crop or plant dry-matter accumulation *in the field* (see Passioura, 1973). In subsequent chapters three of these determinants will be examined in greater detail (ε, J and η), and it will be shown how they can be described in terms of physiological processes occurring at lower levels of biological organization. If eqn (1.4) were to be used as a basis for a predictive, mathematical model of crop yield, such a model could be constructed from an empirical description of the dependence of each of these determinants on the crop environment and need not depend upon a mechanistic understanding of them. If such an empirical, predictive model of crop growth and performance were developed, the empirical description of any of the physiological determinants of growth could be replaced by a mechanistic description (model) as our knowledge of the physiological and environmental effectors of the determinant increased.

It can be argued that, since, in many cropping environments, both the rate and amount of dry-matter production are primarily determined by the availability of water or mineral nutrients, or both (or by pests and diseases), eqns. (1.2)–(1.4) are inappropriate for use in these situations. This argument can be countered on two grounds.

Firstly, environmental factors other than light are implicit in the analysis through their effects on the five determinants of growth; ε, J, η, V and t. Indeed, the value of the analysis as it is set up is that it can be extended, either theoretically or through experimental observation and measurement, to describe environmental effectors of each of the five determinants. For example, this type of analysis has been used to enable studies of the effects of soil water deficits on the rate of leaf net photosynthesis to be quantitatively and unambiguously related to the effects of soil water deficits on dry-matter production by whole plants (Fisher and Charles-Edwards, 1982; Fisher *et al.*, 1981).

Secondly, we can only compare the performances of crops grown at different geographic locations, or at the same location but at different times of the year, if there is some common, well-defined basis for the comparison. Whereas we can modify the rooting environment by applying fertilizers or irrigation, we cannot readily modify the aerial environment. In the first instance, therefore, comparisons of the performances of different crops need to be made in the absence of water or mineral nutrient deficiencies (and pests and diseases), and differences in the aerial environments of crops need to be explicitly accounted for. It is no accident that in recent years attention has become focussed on the efficiency with which plants use light energy in the production of new dry matter (Cooper, 1970; Monteith, 1977; Warren Wilson, 1971). However, an important part of our research must be directed at establishing the quantitative effects of water and mineral nutrient deficiencies on the five logically deduced determinants of crop growth, and extending our analysis of these determinants to take explicit account of these effects.

1.3 Constraints

Synopsis: There are real and inescapable constraints upon the use of the analysis to investigate crop growth. We can usually measure only the *net* rate of crop dry-matter production, and unless we can independently measure or infer either the *gross* rate of dry-matter production or the rate of dry-matter loss, we cannot make unambiguous estimates of ε, the efficiency of light use by the crop. Differences in ε between crops may be real differences, or they may be artifacts due to mistaken estimates of the *gross* rates of crop dry-matter production.

Although the analysis represents a valid, general description of plant growth, its application to particular sets of crop growth data requires assumptions to be made about factors such as dry-matter partitioning. It is difficult to envisage a definitive mathematical growth equation, based on the analysis, because of the great differences in growth strategies of different crops.

Three of the identified physiological determinants of growth can be analysed separately in terms of more basic physiological processes. Because of this, the analysis can provide a useful interface between these more basic areas of plant physiological research and the more applied area of agricultural research.

It is important that we are aware of both practical and theoretical constraints upon our analysis. Whilst the analysis may be based on logical deduction and be applicable to all crops, it still requires measurements of specific characters of those crops to be made. If these characters cannot be measured accurately, the deficiency in our experimental abilities necessarily imposes constraints upon the application of the analysis to the crop growth data.

Whereas we can readily estimate the *gross* amount of new dry matter produced by a crop by dividing the amount of light energy absorbed by it by the calorific content of plant dry matter (for an example see the calculations of Loomis and Williams, 1963), we need to know how the rate of loss of dry matter, V, and thence the amount lost, depends upon both the environment and the physiological state of the crop before we can attempt a similar calculation of the *net* amount of dry matter produced. Conversely, since we can only measure the *net* amount of dry matter present at any moment in time, we need to know something about V in order to estimate both ε and η from our measurements with any confidence. For example, Natarajan and Willey (1980a, 1980b) have linearly regressed the dry-matter yields of sorghum and pigeon pea crops on the cumulated amounts of light energy intercepted by each crop. The slopes of their regressions provide estimates of ε^*, the light-use efficiency in the *net* production of dry matter, for each crop. For the sorghum crop, ε^* was about 2.9 μg (dry matter) J^{-1}, and for the pigeon pea crop it was about 0.9 μg (dry matter) J^{-1}. They remark that the difference between the efficiencies of the two crops probably reflects the differences due to their different photosynthetic pathways (the C_4 photosynthetic pathway for sorghum and the C_3 photosynthetic pathway for pigeon pea). However, their data for pigeon pea seem markedly curvilinear. If we re-write eqn (1.2) in the differential equation form (where we have changed the independent variable t for E, the cumulated intercepted light energy, with $E = \int_0^t J \, dt$)

$$dW/dE = \varepsilon(1 - \gamma W) \tag{1.5}$$

where ε is the efficiency with which *new* dry matter is produced and γ is a constant describing dry-matter loss by the crop. We can integrate over the time interval $0-t$ days to obtain the expression

$$W = [1 - \exp(-\varepsilon\gamma E)]/\gamma. \tag{1.6}$$

The product $\varepsilon\gamma W$ is equivalent to V in eqn (1.2). Note that if we write the coefficient $\gamma = 0$, integration of eqn (1.5) leads to the simple solution

$$W = \varepsilon E \tag{1.7}$$

which is the linear regression of W on E used by the authors. Their data for the pigeon pea crop seem to be as well described by eqn (1.6) as by the simple linear regression (see Fig. 1.1), and they could be described by eqn (1.6) with values for ε of 2.9 and 2.2 μg (dry matter) J^{-1}, and for γ of 0 and 1.5 $\times$ 10^{-3} $m^2 g^{-1}$ for the sorghum and pigeon pea crops respectively. The large difference between the estimates of ε^* for each of the two crops has now almost disappeared in our calculation of ε, and it could be argued that the main difference between them lies in their different values for γ and is not attributable to their different photosynthetic pathways. The analysis is simple and subject to some criticism, but it serves here to make the important point that we must exercise caution whenever we attempt to infer absolute quantities from measurements of *net* amounts of dry matter, or *net* rates of change of the amount of dry matter.

It is difficult to envisage a definitive analysis of crop growth data that could be applied directly to all crops. Crops differ in their growth habits, the nature of their economic, harvestable parts, and in their more basic physiological properties. For example, some plants which are botanically indeterminate, that is their terminal apical meristem retains its capacity for vegetative growth during flowering, behave in an agriculturally determined way. The fibre plant kenaf (*Hibiscus cannabinus*) is botanically indeterminate, but whilst it is flowering and producing seed it effectively ceases vegetative growth. Stem extension growth is curtailed, and the production of new leaves ceases until the reproductive phase of growth is over. In direct contrast, the tomato plant, which is botanically determinate, continues vegetative growth during fruit production. Although the fruit is produced on the terminal apical meristem, an axillary meristem develops vegetatively to produce new stem extension growth and leaves.

Equations (1.2)–(1.4) identify five physiological determinants of both the rate and amount of crop growth and are applicable to both determinate and indeterminate crops. The problem facing agricultural research scientists is to establish the environmental and physiological effectors of each of these five determinants of growth, and the exact role of each one of them in determining crop yield in different cropping situations. Three of them, the light-utilization efficiency, ε, the amount of light intercepted, J, and the partition coefficient, η, can be related directly to our more fundamental, physiological understanding of plant and crop processes. For example, the efficiency with which a crop uses intercepted light energy in the production of new dry matter can be related in some detail to precise physiological properties of

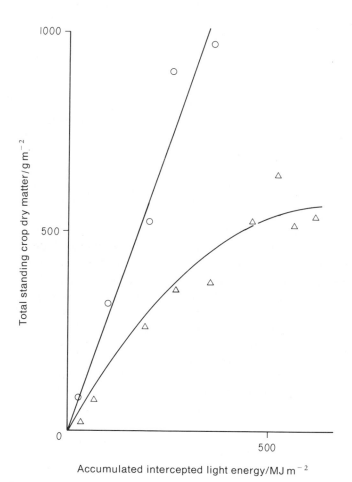

Fig. 1.1. The relationships between total standing crop dry weight and accumulated intercepted light energy for sorghum (○) and pigeon pea (△) crops. The solid lines are growth functions described by eqn (1.6) (see text) (data from Natarajan and Willey, 1980a, 1980b).

plants and crops, properties such as rates of light-saturated leaf photosynthesis and the canopy light extinction coefficients, providing a way of quantitatively integrating this information into agricultural studies. The crop light-use efficiency can also be readily and directly estimated from crop growth data and related, albeit empirically, to changes in the cropping regimen. The analysis therefore provides a convenient interface between the more basic physiological studies of plant processes and the more practical, applied studies of the agricultural sciences.

2

Dry-Matter Production

2.1 An analysis of plant growth data

Synopsis: The simple logical analysis of plant/crop growth data can be applied directly to a variety of published data. It has been used, for example, to identify quantitatively the simultaneous effects of induced soil water deficits on the light-utilization efficiency, ε, and rate of dry-matter loss, V, of plants of the forage legume *Macroptilium atropurpureum*. There is a large body of published data on whole-plant growth studies conducted on seedlings grown in controlled environment facilities. The analysis can be applied directly to much of these data. The roles of basic physiological processes in determining the rate and extent of plant growth can be examined directly at the whole-plant level by investigating the effects of the plant's growing environment on each of the five physiological determinants of growth. For instance, the effects on growth of suppressing the photorespiratory pathway in C_3 plants can be assessed.

It is one thing to develop a logical analysis of plant or crop growth, and it is quite another to apply it to real data. It is often much easier to develop and apply a new analysis for the *specific* case and use the experience and expertise gained by this application to generalize the analysis. The simplest crop system is a single, well-spaced plant, growing free from competition

13

with its neighbours. Consider a single, spaced plant growing in a controlled environment room under a constant and reproducable light level and temperature. If we ignore any shading of the plant by its neighbours or self-shading by its own leaves, we can assume that the amount of light intercepted by the plant is directly proportional to its leaf area projected downward onto a horizontal surface beneath it. If the daily light integral (light energy incident per unit ground area) under which the plant is growing is denoted by S, this assumption allows us to write

$$J = bAS \tag{2.1}$$

where J is the amount of light energy intercepted by the plant each day, A is the total leaf area of the plant and b is a proportionality constant. We can define the projected leaf area of the plant, L, as

$$L = bA. \tag{2.2}$$

(The product bA is the leaf area of the plant projected downward onto a horizontal surface beneath the plant.) We can now make the not unreasonable assumption that the loss of plant dry matter by a spaced seedling is primarily confined to the loss of leaf tissues, and if we then define a leaf abscission constant, γ, such that the daily rate of loss of leaf tissue is proportional to the product γW_L, the daily net growth rate of the plant, dW/dt, can be written as

$$dW/dt = \varepsilon J(1 - \gamma W_L) \tag{2.3}$$

where W_L is the standing leaf dry weight at any particular time and ε the efficiency with which the plant uses intercepted light energy in the production of new dry matter. Equation (2.3) can be compared directly with eqn (1.2) in the previous chapter.

Now, the total leaf area of the plant, A, can be deduced from the leaf dry weight, W_L, using the relationship

$$A = s_A W_L \tag{2.4}$$

where s_A is the specific leaf area (leaf area per unit leaf dry weight). If we use eqns (2.1) and (2.4) to substitute for J in eqn (2.3) we obtain

$$dW/dt = \varepsilon b s_A W_L S(1 - \gamma W_L) \tag{2.5}$$

an equation which relates the net growth rate of the plant to both defined physiological and environmental parameters, and the state of the plant (W_L).

It is convenient, for the moment, if we now restrict our analysis to the vegetative growth phase of the plant. If we denote the proportions of the newly acquired dry matter partitioned to the leaves, stems and roots by η_L, η_S and η_R, we know that during vegetative growth

$$\eta_L + \eta_S + \eta_R = 1. \tag{2.6}$$

Equation (2.6) is a formal statement of the trivial observation that, if a plant has only leaves, stems and roots, all new dry matter must be partitioned to either leaves, stems or roots, and the sum of the dry-matter increment of all three will be equal to the total increment in new plant dry matter. We can then use eqns (2.5) and (2.6) to write the net rates of leaf, stem and root dry-matter growth during the vegetative growth phase as

$$dW_L/dt = \eta_L \varepsilon bs_A W_L S(1 - \gamma W_L) \tag{2.7a}$$

$$dW_S/dt = \eta_S \varepsilon bs_A W_L S \tag{2.7b}$$

$$dW_R/dt = \eta_R \varepsilon bs_A W_L S \tag{2.7c}$$

where the subscripts L, S and R denote leaf, stem and root tissues respectively. The products $\eta_L \varepsilon bs_A S$, $\eta_S \varepsilon bs_A S$ and $\eta_R \varepsilon bs_A S$ are conveniently replaced by μ_L^*, μ_S^* and μ_R^*, which represent "modified" leaf, stem and root specific growth rates. (They are called "modified" specific growth rates because they represent growth rates per unit of *leaf* dry weight.) Equations (2.7a,b,c) can then be integrated over the time interval 0–t days. Using the subscripts L0, S0 and R0 to denote the initial leaf, stem and root dry weights at time $t = 0$, we obtain on integration the three relationships

$$W_L = W_{L0} \exp(\mu_L^* t)/\{1 - \gamma W_{L0}[1 - \exp(\mu_L^* t)]\} \tag{2.8a}$$

$$W_S = W_{S0} + \mu_S^* t/\gamma + (\mu_S^*/\mu_L^* \gamma) \ln(W_{L0}/W_L) \tag{2.8b}$$

$$W_R = W_{R0} + \mu_R^* t/\gamma + (\mu_R^*/\mu_L^* \gamma) \ln(W_{L0}/W_L). \tag{2.8c}$$

Equations (2.8a,b,c) can be used directly to analyse sequential harvest data for spaced plants, and their use is well illustrated by the following example.

Single, spaced plants of the legume *Macroptilium atropurpureum* (cv. siratro) were grown in a controlled environment room. Whilst half of the plants were adequately watered throughout their growth, the remainder were subject to seven successive cycles of drying and re-watering; cyclic water-stressing. Equations (2.8a,b,c) were fitted to experimental harvest data obtained for both the watered and the stressed plants. These harvest data, together with the fitted growth curves for leaf, stem and root fractions of the watered and stressed plants, are shown in Fig. 2.1, and the estimates of the four growth parameters μ_L^*, μ_S^*, μ_R^* and γ for each set of plants are given in Table 2.1. The depression in the total plant "modified" specific growth rate μ_T^* ($= \mu_L^* + \mu_S^* + \mu_R^*$), due to the water-stress treatment corresponds to a 27% reduction in the rate of photosynthesis by the stressed plants. Independent measurements of leaf photosynthesis indicated a 29% reduction in the photosynthetic activity of the stressed plants. But note also that the leaf abscission constant, γ, was increased five-fold by the stress treatment. The analysis enables us to attribute the quantitative difference in the growth of the watered and the stressed plants to the separate, but simultaneous, effects of water stress on leaf photosynthesis and leaf abscission. Further, eqns (2.8a,b,c) enable us to explore the relative effects of water stress on photosynthesis and leaf abscission at different stages of plant growth. (These data are reported in more detail in Fisher and Charles-Edwards, 1982.)

When we consider both vegetative and reproductive growth we need to make additional assumptions. Firstly, we have to decide upon the boundary between the vegetative and reproductive growth states: we have to decide at what point in time the plant changes from vegetative to reproductive growth. Secondly, when we have defined the moment of change, we have to make assumptions about the new pattern of dry-matter partitioning during reproductive growth. This will depend upon the plant type—whether it is a determinate plant or an indeterminate one. Some of these problems will be examined in later sections.

When we are dealing with small spaced plants, particularly during the early stages of vegetative growth, we can generally neglect losses of dry matter through leaf abscission, etc. If we do this we can re-write eqn (2.3) in the simple form

$$dW/dt = \varepsilon J. \tag{2.9}$$

Table 2.1

Estimated leaf, stem and root "modified" specific growth rates (μ_L^*, μ_S^* and μ_R^*) and leaf abscission constant (γ) obtained by fitting eqns (2.8a,b,c) to experimental harvest data for watered and stressed siratro plants. Values in brackets are the approximate 95% confidence intervals.

	$\mu_L^*/10^{-2}\,\mathrm{d}^{-1}$	$\mu_S^*/10^{-2}\,\mathrm{d}^{-1}$	$\mu_R^*/10^{-2}\,\mathrm{d}^{-1}$	γ/g^{-1}
Watered	8.4 (1.4)	7.3 (0.4)	1.8 (0.2)	0.6 (0.1)
Stressed	5.4 (0.5)	5.1 (0.4)	1.6 (0.2)	2.5 (0.6)

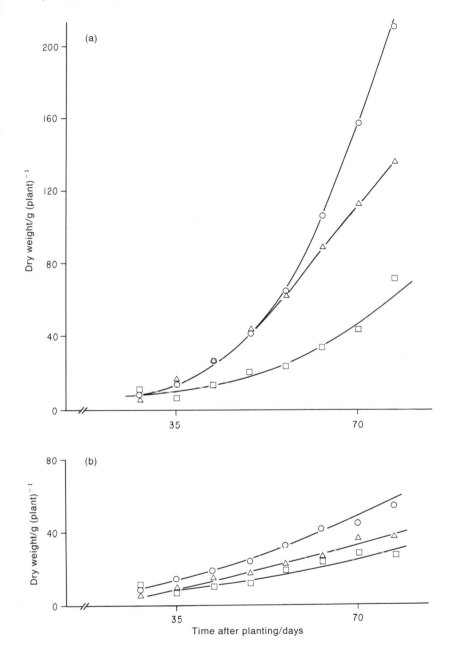

Fig. 2.1. Harvest data and fitted growth functions (eqns (2.8a,b,c)) for the stem (○), leaf (△) and root (□) component dry weights of (a) watered and (b) stressed plants of the forage legume siratro grown in a controlled environment room (data from Fisher and Charles-Edwards, 1982).

Over any time interval Δt days, the amount of light energy intercepted by the plant, E, can be calculated as

$$E = bS \int_0^{\Delta t} A\, dt \tag{2.10}$$

(note that we are assuming that both b and S remain constant over the time interval). If the measured increment in plant dry weight over the same time interval is ΔW, the product $b\varepsilon$ can be simply estimated from the ratio $\Delta W/E$. We can use eqn (2.10) to obtain this ratio as

$$b\varepsilon = \Delta W/S \int_0^{\Delta t} A\, dt. \tag{2.11}$$

We know that during the early stages of growth both the dry weight and leaf area of the plant increase approximately exponentially and these increases can be described by the simple empirical relationships

$$W = W_0 \exp(\mu t) \tag{2.12}$$

and

$$A = A_0 \exp(\mu_A t) \tag{2.13}$$

where t is the time interval over which we are interested in the plants, and μ_A is their specific leaf area growth rate. We can then obtain from eqn (2.12) the increment in plant dry weight, ΔW, over the time interval Δt, as

$$\Delta W = W - W_0 = W_0[\exp(\mu \Delta t) - 1]. \tag{2.14}$$

Similarly, from eqn (2.13), we can obtain the integral of the plant's leaf area as

$$\int_0^{\Delta t} A\, dt = A_0[\exp(\mu_A \Delta t) - 1]/\mu_A. \tag{2.15}$$

If we now use eqns (2.14) and (2.15) to substitute for ΔW and $\int A\, dt$ in eqn (2.11) we obtain the expression for $b\varepsilon$,

$$b\varepsilon = \mu_A W_0[\exp(\mu \Delta t) - 1]/A_0[\exp(\mu_A \Delta t) - 1]. \tag{2.16}$$

Often the published accounts of plant growth experiments record only the initial measurements of the plant dry weight and leaf area, W_0 and A_0, and measurements of W and A made on only one occasion later, Δt days after the commencement of the experiment. In these cases we can use eqns (2.12) and (2.13) to substitute for μ and μ_A in eqn (2.16), thereby obtaining the expression

$$b\varepsilon = (W - W_0)\ln(A/A_0)/(A - A_0)\Delta t. \tag{2.17}$$

There is a considerable body of published data on the growth of spaced plants in controlled environment facilities, and eqn (2.17) can be used to estimate the product $b\varepsilon$ from many of these data sets. Although we cannot be sure whether genetic of environmental differences in the product $b\varepsilon$ are due to differences in b, ε, or to differences in both, the comparison of estimates of the product $b\varepsilon$ can be rewarding. I will call the product $b\varepsilon$ the modified plant light-utilization efficiency.

Imai and Murata (1979a, 1979b) have published a variety of plant growth data on the effects of different light levels, carbon dioxide and oxygen concentrations and temperatures, during growth, on the dry-matter production of young plants of four crop species. Two of the species they examined exhibit the C_3 carbon pathway for photosynthesis (rice and soybean) whilst the other two exhibit the C_4 carbon pathway (millet and maize). Estimates of $b\varepsilon$ can be readily made from their data using eqn (2.17), and these estimates are given in Table 2.2. The four species appear to respond to changes in the light level and temperature during growth in a similar way. The product $b\varepsilon$ for these plants generally decreased with increased growth light level and increased with an increased growth temperature.

Unless we know the ratio of the actual leaf area of the plant to its leaf area projected onto a horizontal surface, and can calculate b, we cannot be sure whether the changes in the product $b\varepsilon$ reflect changes in b, ε or both. However, a subset of the data shown in Table 2.2 are worth examining in more detail. At normal ambient carbon dioxide concentrations (300 vpm) there is a substantial rate of photorespiration by the leaves of C_3 plants. By way of contrast, photorespiration is absent in the leaves of C_4 plants. Photorespiration arises from the photo-oxidation of ribulose-1,5-bisphosphate, and is inhibited if the ambient oxygen concentration in the air surrounding the leaf of a C_3 plant is reduced. There is a concomittant enhancement of the net rate of leaf photosynthesis. Reduced ambient oxygen concentrations have little or no effect on the net rate of photosynthesis by leaves of C_4 plants. Rice and soybean are C_3 plants, whilst millet and maize are C_4 plants. Their "modified" light-use efficiencies when grown at 21% and 6% ambient oxygen concentrations, and 300 vpm ambient carbon dioxide concentration can be abstracted from Table 2.2. They are shown in Table 2.3. Whereas the light-utilization efficiencies of the two C_3 species show marked enhancement when the ambient oxygen concentration is reduced, there is little effect on $b\varepsilon$ for the two C_4 species. At high ambient carbon dioxide concentrations

(1000 vpm), photorespiration in the leaves of C_3 plants is almost entirely supressed, and reducing the ambient oxygen concentration has little effect on the net rate of leaf photosynthesis. The light-use efficiencies of plants grown at 1000 vpm ambient carbon dioxide concentration are insensitive to different ambient oxygen concentrations during growth (see Table 2.3). These plant growth data are therefore in accord with our more basic knowledge and understanding of the photosynthetic properties of C_3 and C_4 plants.

Similarly, estimates of $b\varepsilon$ for a number of C_3 and C_4 plant species grown out-of-doors can be made from the published data of Vong and Murata (1978). These estimates are shown in Table 2.4, and it can be noted that there is no clear distinction in the values of $b\varepsilon$, at any particular temperature, between the C_3 and C_4 plant species. It may be that the lack of difference in the values of $b\varepsilon$ is due to differences in the display of the leaves on the different plant types, that is because of compensatory changes in b. The importance of the display of the light-intercepting, photosynthetic surface to plant growth is well illustrated in the following example.

Table 2.2

The effects of different light levels, temperatures and ambient carbon dioxide and oxygen concentrations on the modified plant light-utilization efficiencies ($b\varepsilon$) of plants of (1) *Oryza sativa* (rice), (2) *Glycine max* (soybean), (3) *Echinochloa frumentacea* (Japanese millet) and (4) *Zea mays* (maize). Plants were all grown under a 12/12 hour day/night regimen. (Data from Imai and Murata, 1979a, 1979b).

				$b\varepsilon/\mu g$ (dry matter) J^{-1}			
S	T	CO_2	O_2	(1)	(2)	(3)	(4)
4.3	23/20	300	21	1.8	1.6	3.3	1.5
4.3	28/23	300	21	1.6	2.0	3.6	1.7
4.3	23/20	1000	21	2.7	2.2	4.1	1.8
4.3	28/23	1000	21	2.8	3.0	3.4	2.0
7.8	23/20	300	21	1.4	1.0	1.8	0.9
7.8	28/23	300	21	1.8	1.2	2.0	1.0
7.8	23/20	1000	21	2.0	1.4	2.0	1.0
7.8	28/23	1000	21	2.8	2.0	2.4	1.4
7.8	28/23	300	6	2.8	1.6	2.1	1.2
7.8	28/23	1000	6	3.1	2.1	2.3	1.4

S — daily light integral (MJ $m^{-2} d^{-1}$).
T — day/night temperature (°C).
CO_2 — ambient CO_2 concentration (vpm).
O_2 — ambient O_2 concentration (%).

Equations (2.7a,b,c) also tell us that genetic or environmental differences in specific leaf area (s_A) and dry matter partitioning (η) can be as important causes of the differences in plant growth rate as differences in the product $b\varepsilon$. We can estimate the "average" specific leaf areas and leaf partition coefficients of spaced plants during their growth from the equations

$$s_A = A_0[\exp(\mu_A \Delta t) - 1]/W_{L0}[\exp(\mu_L \Delta t) - 1] \qquad (2.18)$$

and

$$\eta_L = W_{L0}[\exp(\mu_L \Delta t) - 1]/W_0[\exp(\mu \Delta t) - 1] \qquad (2.19)$$

(cf. eqns (2.12) and (2.13)). Estimates of $b\varepsilon$, s_A and η_L can be obtained (using eqns (2.16), (2.18) and (2.19)) for nine plant species growing in three contrasting temperature regimens from the data of Potter and Jones (1977). These estimates are given in Table 2.5. The salient feature of these plant growth data is that the differences in both s_A and η_L between the contrasting plant species and growth regimens are as large as the differences in the product $b\varepsilon$. Put simply, these data show us that the growth rate of the plant depends as much on its ability to intercept the incident light energy as it does on its capacity to use that light energy in the production of new dry matter.

These data provide an insight into another major and vexing problem: the unsuspected intercorrelations of plant characters. There are demonstrably large differences in the rate of light-saturated, leaf net photosynthesis within both C_3 and C_4 plant types. These differences are illustrated in Table 2.6, where the two-fold to three-fold differences observed within some major

Table 2.3

Estimates of the modified plant light-utilization efficiency ($b\varepsilon$) made from dry-matter growth data for seedlings of C_3 plants (rice and soybean) and C_4 plants (maize and millet) growth at (a) normal ambient carbon dioxide concentrations (300 vpm) and (b) high ambient carbon dioxide concentrations (1000 vpm) at high and low ambient oxygen concentrations (see Table 2.2).

CO$_2$	O$_2$	Rice	Soybean	Millet	Maize
		$b\varepsilon/\mu g$ (dry matter) J^{-1}			
(a) 300	21%	1.8	1.2	2.0	1.0
	6%	2.8	1.6	2.1	1.2
(b) 1000	21%	2.8	2.0	2.4	1.4
	6%	3.1	2.1	2.3	1.4

crop species are illustrated. The differences have been correlated with geno-typic differences in specific leaf area on a dry-weight basis in alfalfa (Pearce *et al.*, 1969), soybean (Dornhoff and Shibles, 1970) and oats (Criswell and Shibles, 1971), with differences in specific leaf area on a fresh-weight basis in maize (Heichel and Musgrave, 1969) and with differences in leaf thickness in soybean (Dornhoff and Shibles, 1976), ryegrass (Charles-Edwards *et al.*,

Table 2.4

Estimates of the modified plant light-utilization efficiency ($b\varepsilon$) for a number of C_3 and C_4 plant species growing in pots out-of-doors under different mean daily light integrals ($\bar{S}$) and mean daily temperatures ($\bar{T}$). (Derived from the data of Vong and Murata, 1978).

(a) C_3 species

| S/MJ m^{-2} d^{-1} | $\bar{T}$/°C | \multicolumn{7}{c}{$b\varepsilon$/μg (dry matter) J^{-1}} | | | | | | |
		(1)	(2)	(3)	(4)	(5)	(6)	All C_3
2.8	17.0	0.9	0.9	0.6	0.9	0.5	0.5	0.7
3.7	21.5	1.0	1.3	-	1.5	2.2	1.9	1.6
2.9	22.5	2.1	1.7	1.6	1.9	2.6	3.2	2.2
4.1	24.0	1.6	1.3	1.0	1.8	1.8	2.1	1.6
4.8	28.0	2.5	0.9	0.9	1.5	-	2.4	1.6
7.2	30.5	1.3	1.2	0.7	1.1	2.5	2.6	1.6

(b) C_4 species

| S/MJ m^{-2} d^{-1} | $\bar{T}$/°C | \multicolumn{5}{c}{$b\varepsilon$/μg (dry matter) J^{-1}} | | | | |
		(1)	(2)	(3)	(4)	All C_4
2.8	17.0	1.1	0.8	0.6	1.4	1.0
3.7	21.5	1.4	1.8	1.7	2.1	1.8
2.9	22.5	1.7	2.0	1.3	2.4	1.9
4.1	27.0	1.3	1.5	1.7	2.2	1.7
4.8	28.0	1.4	1.4	1.9	1.3	1.7
7.2	30.5	1.0	1.8	1.7	2.3	1.7

C_3 species
  (1) *Triticum aestivum* (wheat)
  (2) *Hordeum vulgare* (barley)
  (3) *Pisum sativum* (pea)
  (4) *Glycine max* (soybean)
  (5) *Oryza sativa* subsp. japonica (rice)
  (6) *O. sativa* subsp. indica (rice)

C_4 species
  (1) *Zea mays* (maize)
  (2) *Sorghum bicolor* (sorghum)
  (3) *Panicum miliaceum* (millet)
  (4) *Echinochloa frumentacea*
    (barnyard millet)

1974) and sugarcane (Irvine, 1967). These correlations suggest that a major source of the genetic variation in the differences lies in the size of the photosynthetic system beneath unit leaf area rather than in the specific activity of that system. The plant light-utilization efficiency, ε, is proportional to the rate of light-saturated photosynthesis (see Chapter 4). If increases in ε are associated with increases in s_A or leaf thickness, we might expect plants with higher light-use efficiencies to have lower rates of leaf-area development, and we might also expect them to intercept less light. Whilst at a particular leaf area plants with a higher light-use efficiency may be more productive, this advantage may be offset by the time it takes them to develop an effective

Table 2.5
Estimates of the modified plant light-utilization efficiency ($b\varepsilon$) (μg J^{-1}), specific leaf area (s_A) (10^{-2} m^2 g^{-1}), and leaf partition coefficient (η_L) for nine plant species grown under three contrasting temperature regimens with a daily light integral of 5.0 MJ m^{-2} during a 14 hour light period. (Derived from the data of Potter and Jones, 1977).

| | Day/night temperature regimen/°C | | | | | | | | |
| | 21/10 | | | 32/21 | | | 38/27 | | |
	$b\varepsilon$	s_A	η_L	$b\varepsilon$	s_A	η_L	$b\varepsilon$	s_A	η_L
(1)	1.9	2.3	0.60	4.0	3.8	0.46	3.2	2.9	0.41
(2)	1.8	1.3	0.62	3.2	2.3	0.53	3.0	2.0	0.57
(3)	2.1	2.2	0.53	2.9	2.9	0.46	3.0	2.4	0.46
(4)	2.9	2.2	0.48	3.5	3.1	0.53	3.4	2.7	0.57
(5)	3.4	1.4	0.56	3.4	2.7	0.57	3.3	2.6	0.57
(6)	2.7	2.1	0.35	2.6	3.3	0.64	3.8	2.6	0.45
(7)	2.9	2.0	0.53	4.1	2.9	0.44	3.4	2.3	0.46
(8)	4.5	2.3	0.26	4.5	3.6	0.45	4.7	3.3	0.42
(9)	3.6	2.7	0.49	5.3	3.5	0.47	7.6	1.6	0.53
All C$_4$	3.3	1.8	0.45	4.6	3.6	0.46	5.2	2.6	0.45
All C$_3$	2.6	1.5	0.51	3.0	2.9	0.53	3.3	2.4	0.51

Crop species (1) *Zea mays* (maize — C$_4$)
(2) *Gossypium hirsutum* (cotton — C$_3$)
(3) *Glycine max* (soybean — C$_3$)
Weed species (4) *Abutilon theophrasti* (velvet leaf — C$_3$)
(5) *Anoda cristata* (spurred anoda — C$_3$)
(6) *Sida spinosa* (prickly sida — C$_3$)
(7) *Xanthium sylvanicum* (common cocklebur — C$_3$)
(8) *Sorghum halepense* (johnson grass — C$_4$)
(9) *Amaranthus retroflexus* (redroot pigweed — C$_4$)

Table 2.6

Some examples of the varietal differences in the rates of light-saturated leaf photosynthesis F_{max} observed for three major crop species.

Crop	$F_{max}/10^{-3}$ g (CO_2) m^{-2} s^{-1}	No. varieties investigated	Reference
Maize	0.8–2.4	27	Heichel and Musgrave (1969)
	1.3–1.6	12	Duncan and Hesketh (1968)
Soybean	0.8–1.2	20	Dornhoff and Shibles (1970)
Oats	0.6–0.9	20	Criswell and Shibles (1971)
	0.7–1.3	20	

light-intercepting canopy. Indeed, a crop whose *early* production is of economic importance may benefit from a rapid rate of leaf-area expansion even if this rapid rate of leaf-area expansion does turn out to be associated with thin leaves and a low light-utilization efficiency (Charles-Edwards, 1978).

2.2   An analysis of crop growth data

Synopsis: The analysis can be extended to deal with cropping systems, and used to obtain a quantitative understanding of the role of each of the five physiological determinants of growth. For example, it can be used to demonstrate that the differences in forage yield of three contrasting ecotypes of the forage legume *Stylosanthes humilis* grown at the same site and same time of year are entirely attributable to differences in their phenologies. In contrast, differences in the seed yields of mungbean crops (*Vigna radiata*) grown at different initial plant densities could be attributed to differences in the length of time taken by the crops to achieve "full ground cover", that is to differences in the amount of light energy intercepted by each crop.

We can use the analysis described in the previous section to examine crop growth data. The main difference is that we are better able to define J, the amount of light energy intercepted by the "closed" crop canopy. For the "closed" crop canopy we can replace eqn (2.1) by the relationship

$$J = [1 - \exp(-kL)]S \tag{2.20}$$

where k is a canopy light extinction coefficient and L is the crop's leaf area index. Equations (2.7a,b,c) can then be re-written as difference equations, using eqn (2.20):

$$\Delta W_L/\Delta t = \eta_L \varepsilon S[1 - \exp(-ks_A W_L)] - \gamma W_L \qquad (2.21a)$$

$$\Delta W_S/\Delta t = \eta_S \varepsilon S[1 - \exp(-ks_A W_L)] \qquad (2.21b)$$

$$\Delta W_R/\Delta t = \eta_R \varepsilon S[1 - \exp(-ks_A W_L)] \qquad (2.21c)$$

where W_L, W_S and W_R become the dry weights per unit ground area of the leaf, stem and root components of the crop. Equations (2.21a,b,c) describe vegetative growth by the crop. If we want to include the reproductive growth phase in our analysis, we need to make some additional assumptions about the partition of newly produced dry matter to the reproductive parts of the crop. For an agriculturally determinate crop, that is one which ceases to produce new vegetative extension growth and new leaves at, or soon after, flowering, we can assume that a proportion, β, of the dry matter previously partitioned to the leaves, stems and roots is partitioned to the reproductive parts and the remainder, $(1 - \beta)$, is partitioned to the roots or to a storage organ. Equations (2.21a,b,c) can then be written as

$$\Delta W_L/\Delta t = -\gamma W_L \qquad (2.22a)$$

$$\Delta W_S/\Delta t - 0 \qquad (2.22b)$$

$$\Delta W_F/\Delta t = \beta \varepsilon S[1 - \exp(-ks_A W_L)] \qquad (2.22c)$$

$$\Delta W_R/\Delta t = (1 - \beta)\varepsilon S[1 - \exp(-ks_A W_L)] \qquad (2.22d)$$

when $t > t_f$; where t_f is the median flowering date. Equations (2.21) and (2.22) can be numerically integrated and fitted to experimental crop harvest data. Variations on them have been used to analyse field growth data for crops of the forage legume *Stylosanthes humilis* (Fisher *et al.*, 1980) and the grain legumes *Vigna radiata* and *Vigna mungo* (Muchow and Charles-Edwards, 1982a) grown in tropical northern Australia. The product $\eta_L \varepsilon S$ represents the maximum rate of leaf dry-matter production by the crop, when it is intercepting all the light energy incident upon it. Similarly, $\eta_S \varepsilon S$ is the maximum rate of stem dry-matter production. As rates of dry-matter production per unit ground area they will have dimensions of $g\,m^{-2}\,d^{-1}$ or $t\,ha^{-1}\,d^{-1}$ (note that $1\,t\,ha^{-1}\,d^{-1} \equiv 100\,g\,m^{-2}\,d^{-1}$). Estimates of $\eta_L \varepsilon S$ and $\eta_S \varepsilon S$ obtained from the analysis of the growth of these crops are given in Table 2.7. The average daily light integrals during the growth periods of all three crops were about 9.5 MJ m^{-2}. The mungbean crops were grown when the average air temperature was about 29°C and the stylo crops were grown with the air temperature averaging about 27.5°C. Whereas β was estimated at 0.69 for the mungbean crops, it was estimated to be only 0.38 for the stylo crops.

Often, as with the crops described above, only the above-ground parts of a crop are harvested. If we denote the efficiency of use of light in the production of new above-ground dry matter by ε_T, we can write it as

$$\varepsilon_T = (\eta_L + \eta_S)\varepsilon. \tag{2.23}$$

The efficiencies of light use in the production of new above-ground dry matter for the crops shown in Table 2.7 can be compared with those for the above-ground parts of a C_4 grain sorghum crop, during different growth stages, grown also in northern tropical Australia. This comparison is made in Table 2.8. Whilst only the production of above-ground parts is being compared, there is a suggestion in these data that the efficiency of the C_4 sorghum crop is about 20% greater than those of the C_3 crops when grown at similar mean air temperatures.

The data for the sorghum crop were not obtained by fitting eqns (2.21) and (2.22) to sequential harvest data. They were directly estimated from the dry-matter harvests of above-ground crop parts at each of the four growth stages: initiation, boot, anthesis and maturity. Loss of leaf material was explicitly accounted for in these calculations by assuming that until anthesis the partition of new dry matter to the leaves was the same throughout growth. The partition coefficient was calculated as the ratio of the net standing leaf to stem dry weight at the first harvest. This enabled the gross increment in new leaf dry weight at any time during growth, $(\Delta W_L)_{\text{gross}}$, to be calculated from

$$(\Delta W_L)_{\text{gross}} = \Delta W_S LSR \tag{2.24}$$

Table 2.7

Estimates of $\eta_L \varepsilon S$ and $\eta_S \varepsilon S$ obtained by fitting eqns (2.21) and (2.22) to field data for crops of *S. humilis*, *V. radiata* and *V. mungo* grown in northern tropical Australia.

Crop	$\eta_L \varepsilon S / \text{t ha}^{-1} \text{d}^{-1}$	$\eta_S \varepsilon S / \text{t ha}^{-1} \text{d}^{-1}$
S. humilis	9.1	6.9
V. radiata	11.6	8.6
V. mungo	11.3	9.7

where *LSR* is the ratio of the standing leaf and stem weights at the first harvest. This method of calculation of the gross increment in leaf dry matter makes the assumption that there are no ontogenetic changes in the partitioning of new dry matter between the different plant parts during vegetative growth. The assumption is implicit in the "fitting" of eqns (2.21) and (2.22) to crop growth data, using sophisticated computer techniques. This *simple* "long-hand" method of analysing crop data, based on only two or three harvests of crop dry matter, should therefore produce results that are quite comparable with the computer-aided analysis of extensive sets of crop dry-matter data. After anthesis it was assumed that no new leaf dry matter was produced, and that the gross increment in above-ground dry matter was the sum of the increments in stem and grain dry weight. These estimates of ε_T for the mungbean and sorghum are usefully compared with those for chickpea and sorghum ($\varepsilon = 2.2$ and $2.9 \ \mu g \ J^{-1}$) reported in the previous chapter.

The analysis of crop growth data in this way allows us to distinguish effects of both the cropping environment and the crop-management strategy on the five main physiological determinants of crop growth described in the previous chapter. For example, differences in both the forage and seed yields of three contrasting ecotypes of the forage legume *Stylosanthes humilis*, grown at different times of year in northern Australia, were entirely attributable to differences in the phenology of the different crops, that is the lengths of their vegetative and reproductive growth phases. One set of these data is illustrated in Fig. 2.2. The fitted growth curves for the early, mid-season and late flowering ecotypes of *S. humilis*, sown on the same day in November, are shown. It is clear from the fitted growth curves that the three ecotypes grew in identical manner, and that the only difference between them was their times of flowering. Forage production (standing leaf and stem dry matter)

Table 2.8

The efficiencies of new above-ground dry-matter production (ε_T) for four field crops growing in northern tropical Australia at different mean air temperatures ($\bar{T}$). (Data on *S. bicolor* was provided by R. Muchow).

Crop	$\varepsilon_T/\mu g \ J^{-1}$	$\bar{T}/°C$	Growth stage
S. humilis	1.7	27.5	emergence → maturity
V. radiata	2.1	29.0	emergence → maturity
V. mungo	2.2	29.0	emergence → maturity
S. bicolor	2.1	21.5	initiation → boot
S. bicolor	2.4	23.5	boot → anthesis
S. bicolor	2.5	27.0	anthesis → maturity

was least for the early flowering ecotype and largest for the late flowering one. Conversely, seed production was greatest for the early flowering ecotype and least for the late flowering one. The analysis of this data is reported more fully in Fisher *et al.* (1980).

By way of contrast, an analysis of the growth of mungbean crops, *Vigna radiata* (cv. Berken), grown at different plant densities has indicated that the differences in total above-ground dry matter at flowering, and subsequent differences in seed yield, were almost entirely attributable to differences in the lengths of time taken by the crops to achieve "full ground cover", that is the total amount of light energy intercepted by each crop during its vegetative growth period (Muchow and Charles-Edwards, 1982a).

Whereas it might be argued that these effects would have been obvious at the outset, it was not obvious that there would be no major compensatory

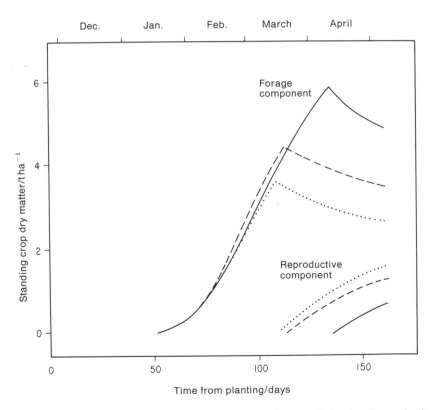

Fig. 2.2. Fitted growth curves for the forage components (leaves and stems) and reproductive component of three ecotypes of *S. humilis* sown at the same time of year. Late flowering ecotype (———); mid-flowering ecotype (– – –); early flowering ecotype (· · · ·).

effects on the other determinants of growth. The analysis can help us to quantitatively and unambiguously identify the main effects of different cropping regimens on the components of dry matter production.

2.3 Crop growth rates

Synopsis: Although there are three distinct physiological determinants of the net rate of dry-matter production by a crop, an examination of the growth rates of contrasting crop species can be very rewarding. There is evidence that very different crops have similar rates of "above-ground" dry-matter production. If they have closed green surfaces, so that they are intercepting all of the light energy incident upon them, the similarity in crop growth rates suggests a similarity in the light-utilization efficiencies of the different crops.

There are many reports of crop growth rates in the literature, and intricate arguments are often woven around unproven or unprovable assumptions in order to make comparisons between the growth rates of contrasting species, or the same species in contrasting growth environments. In particular, differences in the maximum growth rates of crops comprised of C_3 and C_4 plants are often used to support the thesis that C_4 crops are more "efficient" and have a higher growth potential, than C_3 crops. We can use the analysis to write the net growth rate of a crop as

$$\Delta W/\Delta t = \varepsilon J - V \tag{2.25}$$

where V is the rate of loss of crop dry matter. If we are considering only the above-ground parts of the crop we can re-write eqn (2.25) as

$$\Delta W_T/\Delta t = \eta_T \varepsilon J - V_T \tag{2.26}$$

where the subscript T denotes the above-ground parts. Crop growth rates may vary because any, or all, of the determinants ε, J, η_T and V_T (or V) vary and, in general, independent measurements of any of these parameters are rarely reported. The light distribution within a crop canopy can affect the efficiency with which it is able to use that light energy to produce new dry matter. For example, it has been observed that removing side screen-netting from around a block of chrysanthemum plants enhanced their photo-synthetic activity (expressed per unit of ground area), and the subsequent removal of one-fifth of the plants to create a path through the block did not reduce the activity per unit ground area (including pathway area) to less than that of the original screened block of plants (Acock *et al.*, 1978). Whilst these treatments affected J, they also affected ε by changing the light distribution within the crop canopy.

It is worthwhile considering briefly how a change in the light distribution within a crop can affect ε. The rate of leaf net photosynthesis does not increase linearly with increasing light flux densities incident upon the leaf surface. As the light flux density incident upon the leaf increases, the rate of leaf photosynthesis increases less and less quickly, until it reaches a maximum value, the rate of light-saturated leaf net photosynthesis. We can put this in another way. As the light flux density incident upon the leaf surface increases, the efficiency with which it uses that light energy in carbon assimilation (photosynthesis) declines. Clearly, if we compare two canopies absorbing the same amount of light energy, but the "average" light flux density on the illuminated leaves within one canopy is less than the "average" incident upon leaves within the other, the leaves in the first canopy will make more efficient use of the light incident upon them than the leaves in the second canopy. We might therefore expect the first canopy to have a higher light-utilization efficiency than the second. However, if we remove guard plants from around a block of plants we allow light to penetrate the crop canopy from the side. Although the efficiency of the leaves in this block of plants may be reduced, because each leaf is receiving more light, the *total* amount of light energy absorbed by the block may have increased because the crop is now "seeing" light previously absorbed by the guard plants. Some effects of canopy architecture on the crop light-utilization efficiency are discussed in Section 4.4. Comparisons of crop growth rates can only be meaningfully made when the crop area is sufficiently large, or sufficiently well guarded, for us to be able to ignore these edge effects.

When a crop has attained full ground cover, J approaches a maximum value, and provided the seasonal changes in the daily radiation integral are sufficiently small we can write eqn (2.26) as

$$\Delta W_T / \Delta t = \eta_T \varepsilon S = a_T \qquad (2.27)$$

where a_T is the net above-ground growth rate of the crop. Estimates of a_T for a variety of crops grown during the summer months in the Netherlands and the United Kingdom are shown in Table 2.9. These data suggest that the wide, contrasting range of crops differ little in their separate capacities for growth. The C_4 maize crop has a similar growth rate to the C_3 cereal crops (barley, wheat and oats), and we might expect η_T to be similar for all four crops. We would be incautious if we concluded that similarities in a_T for C_3 and C_4 crops grown at the extremes of the latitudinal range for C_4 crops preclude differences at mid-latitudes, but they do suggest that these differences may be small (see also Table 2.8).

Table 2.9

Estimates of the rate of above-ground dry-matter production (a_T) for a variety of crops growing during the summer months in the Netherlands and the United Kingdom.

Crop	$a_T/\mathrm{g\,m^{-2}\,d^{-1}}$	Reference
Spring barley	18	
Spring wheat	18	
Maize	17	Sibma (1968)
Oats	17	
Peas	20	
Peas	19	
Broad bean	14	
Lettuce	19	Greenwood *et al.* (1977)
Spinach	19	
Summer cauliflower	15	
Grass	16	Sibma (1968)

2.4 The duration of growth phases

Synopsis: The duration of production of the harvestable component of the crop has been previously defined as one of the physiological determinants of crop yield. The forage yields of both stylo and brassica crops, grown in Australia and Scotland, increase approximately linearly with the duration of their vegetative growth phases. The seed yields of such diverse crops as kenaf, mungbean and soybean also increase linearly with the duration of the period of reproductive growth. The grain yields of contrasting wheat cultivars can also be written as a function of the duration of the period of grain filling. The problem central to the prediction of the duration of either the vegetative of reproductive growth phases of a crop is the development of a proper understanding of the physiological and environmental effectors of the phenology of the crop.

The duration of crop growth has been defined as one of the physiological determinants of yield. When we examine sets of sequential harvest data, we commonly observe that during the vegetative phase of crop growth there is an initial exponential phase of growth (when the leaf canopy is being established by the crop) but that subsequently the standing crop dry matter increases as a linear function with respect to time (see for example Greenwood *et al.*, 1977). We can simplify this observed relationship by writing the standing crop dry weight, W, at any time t as the simple function of t,

$$W = at + b \qquad (2.28)$$

where a is the maximum crop growth rate (equivalent to a_T) and b is a constant. Data for the forage yields of stylo crops grown in northern Australia (Fisher *et al.*, 1980) and brassica crops grown in south-eastern Scotland (Harper and Compton, 1980) are illustrated in Fig. 2.3, and they demonstrate the applicability of eqn (2.28) to vegetative crop growth quite well. The brassica crops show a distinct curvilinearity in forage dry-matter accumulation after about 100 days of growth. This can be attributed to the marked decline in the daily light integral, S, during the autumn months in Scotland when these particular data were collected. We could re-write eqn (2.28) as

$$W = \varepsilon \bar{Q} \sum S + b \qquad (2.29)$$

where $\bar{Q}$ is the average proportion of the incident light intercepted by the

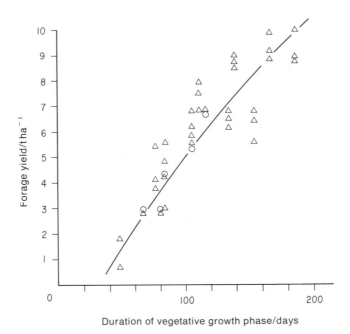

Fig. 2.3. The observed relationship between forage yield and the duration of vegetative growth for field crops of brassica ($\triangle$) grown in Scotland (data from Harper and Compton, 1980) and *Stylosanthes* ($\bigcirc$) grown in northern Australia (data from Fisher *et al.*, 1980).

crop and $\sum S$ is the cumulated daily light integral experienced by the crop (cf. eqns (1.6) and (1.7) in Chapter 1). However, we should also note that in this simple analysis there may be seasonal effects, due to changing temperatures, on ε and $\bar{Q}$. For example, a period of severe frost may cause leaf death, with a consequent reduction in $\bar{Q}$. In general, the longer the phase of vegetative crop growth the greater the yield of the vegetative parts of the crop will be. If we know the seasonal changes in daily light integral and average air temperature, we could use the analysis to predict the potential length of the cropping season for a particular site (see Sections 3.1 and 3.3), and the potential total dry-matter yield of a vegetative crop at that site.

We can also relate the seed yield of both the legume seed crops, mung-bean and soybean, and the non-legume kenaf (*Hibiscus cannabinus*) to the durations of their reproductive growth phases. These relationships, illustrated in Figs. 2.4 and 2.5, could be used as simple predictors of seed yields by these crops. We could re-write eqn (2.29) as

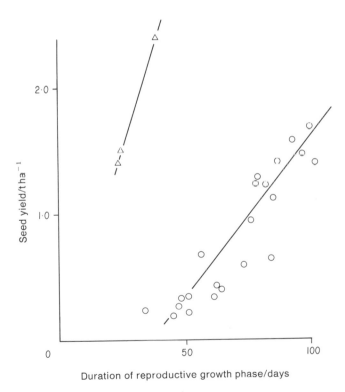

Fig. 2.4. The observed relationship between seed yield and the duration of the reproductive growth phase of mungbean ($\triangle$) (data from Muchow and Charles-Edwards, 1982b) and kenaf ($\bigcirc$) (data from Muchow, Wood and Charles-Edwards, in preparation), grown in northern Australia.

$$W_H = \eta_H \varepsilon \bar{Q} \sum S + b \qquad (2.30)$$

where η_H is the proportion of new dry matter partitioned to the seed, an important harvestable component of these crops. Equation (2.30) can be further simplified by writing $\sum S = \bar{S} \Delta t$, where Δt is the duration of the reproductive growth phase and $\bar{S}$ is the average daily light integral incident upon the crop during reproductive growth, when it becomes

$$W_H = \eta_H \varepsilon \bar{Q} \bar{S} \Delta t + b. \qquad (2.31)$$

If we presume that η_H, ε, $\bar{Q}$ and $\bar{S}$ remained constant for each of the crops shown in Figs. 2.4 and 2.5, eqn (2.31) describes the relationships between W_H and Δt observed for them.

Austin and co-workers (Austin *et al.*, 1980) have examined the improvement in winter wheat yields brought about by the introduction of new cultivars since 1900 in the United Kingdom. They have reported maximum crop leaf-area indices and the duration of grain filling for crops of these

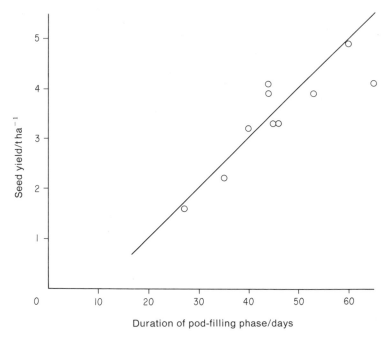

Fig. 2.5. The observed relationship between seed yield and the duration of seed production (pod-filling) for soybean crops grown in eastern Australia (data from Jones and Laing, 1978).

cultivars. If we assume that the canopy light extinction coefficient for all crops was about 0.5 (for most crops it lies between 0.4 and 0.6), from the reported leaf area index for the crop we can use eqns (3.18) and (3.23) to calculate $\bar{Q}$ for each crop. The regression of grain yield, W_H, on the product $\bar{Q}\,\Delta t$ for the twelve cultivars they examined, at two sites, is illustrated in Fig. 2.6. It could be strongly argued that the differences in grain yields of

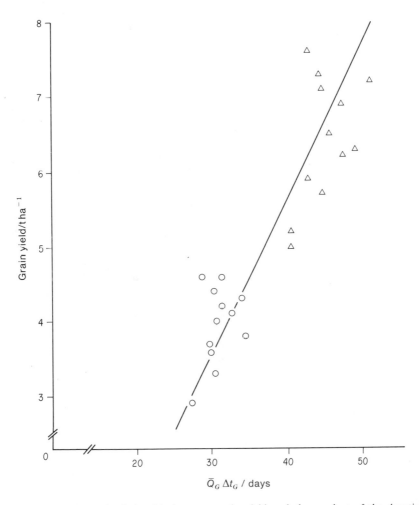

Fig. 2.6. The observed relationship between grain yield and the product of the duration of grain production (Δt_G) and the proportion of incident light energy intercepted during the grain production period ($\bar{Q}_G$) for twelve wheat cultivars grown at two sites at Cambridge, England ($\bigcirc$—data from Patemoster Field; $\triangle$—data from Camp Field).

the twelve cultivars, and the differences in yields between sites, were primarily attributable to the differences in the amounts of light energy intercepted by the crops during grain filling, that is to the product $\overline{Q}\overline{S}\,\Delta t$. The conclusions have immediate implications. Perhaps prolonging of the grain-filling period, either by reducing the rate of leaf senescence or by increasing crop leaf-area index at anthesis (or at the start of grain production) are the major effectors of crop yield, and changing either η_H or ε is less important.

For an agriculturally indeterminate crop, the duration of fruit or seed production will most probably depend upon the partition of assimilate between the reproductive and vegetative plant parts. This partitioning may be determined by other environmental factors, such as temperature, day-length or water availability.

The change from vegetative to reproductive growth will also be determined by environmental factors such as daylength and temperature. There are numerous day/degree, or heat sum, formulae that enable us to predict flowering times for different crops. Our understanding of the effects on, and the changes in, the physiological states of plants as they move from vegetative to reproductive growth is limited, and this is an area of research that is of some importance, particularly for crops grown in areas which have only short, well-defined growing seasons.

Traditionally, physiological studies of the transition from vegetative to reproductive growth have concentrated on those plants which respond rapidly to some external flowering stimulus. Typically, these experiments have identified the proportion of plants in a particular population that have flowered in response to the stimulus, and they provide us with little information on the changing physiological state of the plant as it approaches, and then crosses, the boundary between vegetative and reproductive growth states. The first problem is to define the boundary between states. Thornley highlights this problem in his application of a model of a biochemical switch to the flowering process (see Chapter 2 in Thornley, 1976). In this model he defines the boundary between vegetative and reproductive states by the ratio of the concentrations of a vegetative enzyme and a flowering enzyme. The boundary is defined by a particular numerical value for this ratio, and he is able to demonstrate how suitable stimuli can perturb the system and cause progressive changes in the ratio. At some point the critical ratio is attained, and the system changes between the vegetative and reproductive states. The agronomic definition of the boundary between vegetative and reproductive growth phases is as difficult to define as the physiological one.

Consider a cereal crop such as wheat. In the strict physiological sense it becomes reproductive when the flower is initiated on the terminal apical meristem. However, newly formed leaves continue to expand for some time after flower initiation, and a far greater proportion of dry matter is partitioned to these leaves than to the developing flower. Only at, or about, anthesis does leaf expansion cease. At the level of resolution available with crop dry matter data, the time of anthesis probably reflects the better boundary between the vegetative and reproductive growth phases of the crop. This is a matter of some practical importance and one that is discussed again in Section 6.4.

Barnes (1977) has cogently discussed and examined the influence of the length of the growth period on total crop yield. The mathematics he uses in developing his models for crop yield are complex. However, the salient point of his analysis is that by logically identifying the main determinants of crop growth, and by choosing particular relationships between plant weight and time, yield per unit ground area and plant density, and plant growth rate and plant density, he is able to develop a reasonably general model for crop yield as a function of time. The exercise provides an analogous empirical analysis to the logical physiological analysis described here.

3

Light Interception

3.1 Seasonal and geographical variation in incident radiant energy

Synopsis: Both the daily radiation integral and daylength vary approximately sinusoidally with time of year. We can use sine functions to predict approximate values for them both at any particular time of year and at any location. For many of our needs, approximate estimates of them are adequate.

The annual mean value of the daily radiation integral decreases approximately linearly as a function of increasing latitude. The relative amplitude of the seasonal variation in the daily radiation integral increases approximately linearly with increasing latitude. If we combine these relationships with a sine function describing seasonal variation in the daily radiation integral, we can predict the "average" daily radiation integral on any particular day at a site of known latitude from a knowledge of the site's latitude alone.

The annual mean value of daylength remains constant across latitudes. The relative amplitude of the seasonal variation in the daylength increases approximately exponentially with latitude. We can combine this relationship with a sine function describing seasonal variation in daylength to obtain an estimate of the daylength, at any time of year, for a site at any latitude.

If we develop an analysis of cropping systems based on the efficiency with which the crops use absorbed light energy for the production of new dry matter, it must follow that we need to know the amount of light energy that is incident upon a particular crop and thence available to it for the production of new dry matter. Although we can make direct measurements of the light energy incident upon the crop at particular sites, often we do not have the equipment or the labour resources to measure it at other sites, and we need some simple means of estimating it at them. We do not usually need exact estimates, we only need approximate values. In this section I will describe some very simple relationships which allow us to estimate average daily radiation integrals, allowing for seasonal variation at a particular site, and latitudinal variations between sites.

Whereas there are standard tables predicting seasonal and geographical variation in the radiant solar energy incident at the top of the atmosphere (for example the Smithsonian Meteorological Tables), the radiant energy incident at ground level is not so readily obtained. The path-length of the radiant energy through the earth's atmosphere changes with time of day, season and latitude. The proportion of the radiant energy that is absorbed or scattered during its passage through the atmosphere will depend upon the path-length and the physical characteristics of the atmosphere (the amount of dust in the air, the amount of water vapour, etc.). Variations in the density of dust particles and the density and extent of cloud cover contribute to the day-to-day noise observed in measurements of the daily integral of the downward radiant energy flux density at ground level at any particular site. Data obtained at the Cooper Laboratory in southern Queensland (27°34′S, 152°20′E) during October 1974 illustrate the considerable amount of this noise (Fig. 3.1). In the first instance, we simply want to know long-term, "average" daily values of the incident radiant energy at any particular time of year. As our calculations become more refined, we may also need to be able to predict the distribution of the actual values about this "average" value, and the probability of a crop experiencing a particular amount of incident radiant energy on any given day.

Usually the seasonal variation in the daily integral of the downward radiant energy flux density upon an unobstructed horizontal surface at ground level, R, can be usefully and simply approximated by a sine function of the sort

$$R = R_M + R_D \sin \left[2\pi(t + Z)/365 \right] \tag{3.1a}$$

where R_M is the annual mean value of the integral, R_D is the seasonal amplitude of the integral about the mean, Z is a constant and t is the time variable in days after January 1st. Z depends upon the number of days after January

3. Light Interception

1st that the autumn equinox occurs and has the value of 283 days in the northern hemisphere and 101 days in the southern hemisphere. The application of eqn (3.1a) to data for three sites across the approximate latitudinal range 10°–55° are illustrated in Fig. 3.2.

Seasonal differences in cloud cover (for example summer "wet seasons" in the tropics) affecting the daily radiation integral at ground level may be better accommodated at a particular site by inclusion of a small-phase correction. This apparent "phase shift" due to the summer monsoonal rain is illustrated in the long-term averages for Katherine, NT. As a simple first approximation, the phase shift can be related to the latitude of the site. We could re-write eqn (3.1a) as

$$R = R_M + R_D \sin\{2\pi[t + Z + (55 - N)]/365\} \qquad N < 55 \quad (3.1b)$$

where N is the latitude of the site. The application of eqn (3.1b) to the data for two of the sites shown in Fig. 3.2 is illustrated in Fig. 3.3. It is unlikely that eqn (3.1b) would be globally applicable. It is included to illustrate how the simple sine function could be modified to deal with data from different sites where there are marked and predictable effects of heavy cloud cover causing apparent phase shifts in the seasonal radiation patterns.

The parameters R_M and R_D can be estimated from data for meteorological sites at different latitudes. The latitudinal variation in R_M, and the

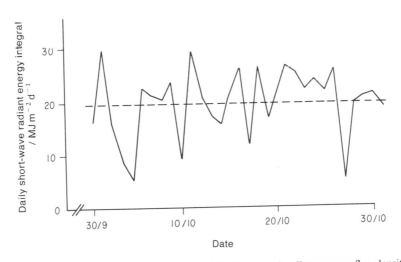

Fig. 3.1. Measurements of the daily integral of the downward radiant energy flux density at the Cooper Laboratory, Queensland (27°34′S, 152°20′E) during October 1974. Average daily integral during the month (––––).

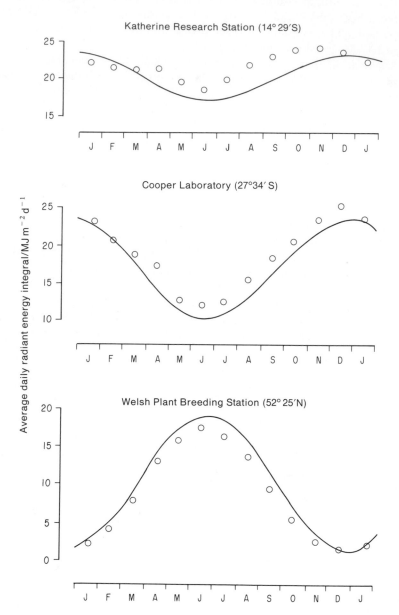

Fig. 3.2. Long-term averages of the monthly means of the daily integral of the downward radiant energy flux density for three sites at different latitudes. The solid lines are obtained by using eqns (3.1a), (3.2) and (3.3) to simulate the data for each site.

relative seasonal amplitude R_D/R_M are given with reasonable accuracy by the simple linear relationships

$$R_M = 24.3 - 0.264N \qquad 10 < N < 55 \qquad (3.2)$$

$$R_D/R_M = 0.0186N - 0.12 \qquad 10 < N < 55 \qquad (3.3)$$

where N is the latitude. The application of eqns (3.2) and (3.3) to data from fifteen sites across the latitudinal range 12°–52° are shown in Fig. 3.4.

If we substitute for R_M and R_D in eqn (3.1a), using eqns (3.2) and (3.3) we obtain the relationship

$$R = (24.3 - 0.264N)\{1 + (0.0186N - 0.12)\sin[2\pi(t + Z)/365]\}$$
$$10 < N < 55 \qquad (3.4)$$

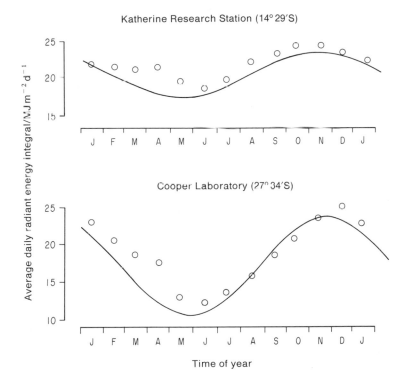

Fig. 3.3. Long-term averages of the monthly means of the daily integral of the downward radiant energy flux density for two sites and predictions of the seasonal variation with explicit allowance made for phase shifts in the seasonal variation (eqn (3.1b)) due to cloud cover.

Equation (3.4) enables us to calculate an approximate value of the average daily radiation integral at ground level at any time of year for any location, from a knowledge of its latitude alone. We must remember that this estimate of R is only an approximate value, but it may be adequate for many of our needs.

We can use a similar mathematical approach to describe the seasonal and geographic variation in daylength. Values for the duration of daylight are tabulated in the Smithsonian Meteorological Tables, and it is easily shown that as a useful and simple approximation the seasonal variation in daylength, h, can be described by

$$h = 4.36 \times 10^4 + h_D \sin\left[2\pi(t + Z)/365\right] \qquad (3.5)$$

where h_D is the seasonal amplitude of the variation in daylength, Z is the

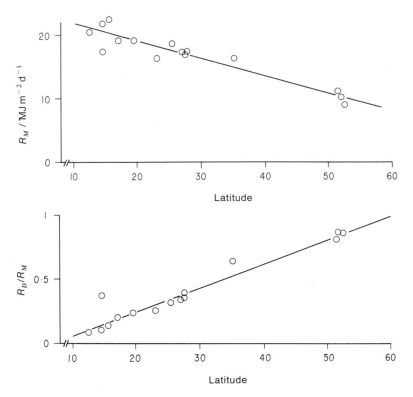

Fig. 3.4. Changes in the annual mean value for the daily integral of the downward radiant energy flux density (R_M) and relative seasonal amplitude of the average daily integral (R_D/R_M) with latitude. The solid lines are given by eqns (3.2) and (3.3).

number of days the autumn equinox follows January 1st and t is the time variable. The annual mean value for the daylength is a little greater than 12 hours (4.32×10^4 seconds) because of the refraction and scattering of the sun's light by the earth's atmosphere. Although the sun may be below the horizon, light is refracted and scattered back down on to the earth's surface by the dense atmosphere. The latitudinal variation in h_D is simply and usefully approximated by

$$h_D = \exp(7.42 + 0.045N) \qquad 10 < N < 55 \qquad (3.6)$$

and this relationship is illustrated in Fig. 3.5.

It needs to be re-emphasized here that eqn (3.4) provides only estimates of the average daily radiation integral for sites at different latitudes. It does not supplant direct measurement: it complements it for those locations where the direct measurement of daily radiation integrals is not possible.

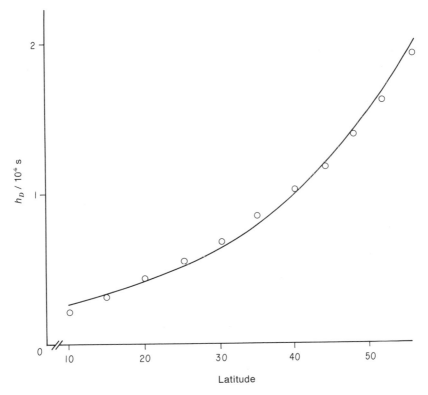

Fig. 3.5. Variation in the amplitude of the seasonal variation in daylength (h_D) with latitude. The solid line is given by eqn (3.6).

3.2   Light quality

Synopsis: Only radiant energy in the 400–700 nm waveband is used by plants for the photosynthetic reduction of gaseous carbon dioxide to the carbohydrate moiety (CH_2O). The energy acquired by an atom or molecule in absorbing one quantum depends upon the frequency of the irradiating light. The energy associated with the 400–700 nm waveband therefore depends upon the spectral composition of the irradiating light. Approximately 45% of the total radiation flux density incident upon an unobstructed surface at ground level is associated with light in the 400–700 nm waveband.

The second law of photochemical equivalence states that any molecule activated by light absorbs one quantum of the light that causes the activation. The energy then acquired by the molecule depends upon the frequency of the light, and is given by Planck's relation $E = hv$, where h is Planck's constant and v is the frequency of the irradiating light. The energy absorbed per mole of substance will be Avogadro's number times hv, or $E = nhv$. This quantity of energy is an Einstein. The energies contained in an Einstein for various wavelengths of radiation are given in Table 3.1.

Measurements of total radiation at the earth's surface include all wavelengths from the shorter, ultraviolet wavelengths to the longer, infrared wavelengths. Measurements of the ratio of the light energy (400–700 nm)

Table 3.1

The energies contained in an Einstein for different wavelengths of radiation.

Wavelength/nm	Spectral range	Energy per Einstein/MJ
100	ultraviolet	1.195
200	ultraviolet	0.598
300	ultraviolet	0.398
400	visible (violet)	0.300
500	visible (blue/green)	0.273
590	visible (yellow)	0.208
650	visible (orange)	0.184
750	visible (red)	0.159
800	infrared	0.149
900	infrared	0.133
1000	infrared	

to the total radiant energy incident upon a horizontal, unobstructed surface within a glasshouse in southern England indicate a negative correlation between this ratio and the total incident radiant energy flux density (Acock *et al.*, 1978). In the early morning and late evening, when the sun angle was low, approximately 63% of the incident radiant energy flux density was in the 400–700 nm waveband. At around noon, when the sun was near its zenith, the ratio had decreased to 54%, although the total incident flux density had increased twenty-fold from about 20 W m^{-2} to around 400 W m^{-2}.

Variation in the annual integral of the incident light flux density under conditions of mean cloudiness have been tabulated by Warren Wilson for different ranges of latitude (see Warren Wilson, 1971, Table 9). If we make the assumption that 45% of the total incident radiant energy flux incident at ground level is in the 400–700 nm waveband, values of the annual integral of the incident light flux density for the sites used in Fig. 3.4 can be compared with Warren Wilson's tabulated values. This is done in Fig. 3.6. The estimate

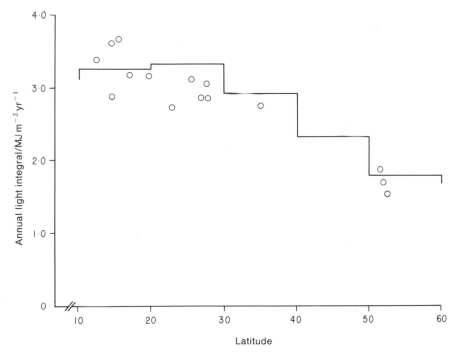

Fig. 3.6. A comparison of tabulated values of the annual mean value for the daily integral of the downward light flux density between different latitudes (Warren Wilson, 1971, Table 9) and estimates (see text) for fifteen sites at different latitudes.

45% for the ratio is one that is commonly used. The higher values of 54–63% found under glass may reflect the different transmission coefficients through glass of radiant energy at different wavelengths. The changing spectral composition of the radiant energy flux with time of day will also reflect the changes in path-length of the radiation through the earth's atmosphere with time of day. The transmission and scattering of radiation through and in the atmosphere will depend upon the wavelength of the radiation.

3.3   Seasonal and geographical variation in temperature

Synopsis: Seasonal variation in the average daily air temperature can be described by a simple sine function. The annual mean value of the average daily air temperature decreases approximately linearly with increasing latitudes. The relative amplitude of the seasonal variation in the average daily air temperature increases approximately linearly with latitude.

Seasonal changes in air temperature are generally out of phase with changes in the incident radiation integral. If we define the average daily air temperature, $\bar{T}$, as the simple mean of the maximum and minimum air temperatures, we can write $\bar{T}$ as

$$\bar{T} = \bar{T}_M + \bar{T}_D \sin\left[2\pi(t + Z - N/2)/365\right] \tag{3.7}$$

where $\bar{T}_M$ is the mean annual temperature, $\bar{T}_D$ is the seasonal amplitude of the variation, Z is as defined in Section 3.1, N is the latitude of the location and t is the time variable in days after January 1st. Equation (3.7) is used to describe data for three sites across the latitudinal range $10°–55°$ in Fig. 3.7. As the latitude of the site increases, the lag in phase between the seasonal changes in radiation and temperature also increases. It is a fortunate and simple coincidence that, as a useful first approximation, this lag can be taken as being directly proportional to the latitude, N, expressed in units of days.

Changes in the mean annual temperature, $\bar{T}_M$, and the relative seasonal amplitude in temperature, $\bar{T}_D/\bar{T}_M$, are usefully and simply approximated by the relationships

$$\bar{T}_M = 32.5 - 0.45N \qquad 10 < N < 55 \tag{3.8}$$

$$\bar{T}_D/\bar{T}_M = 0.015N - 0.10 \qquad 10 < N < 55 \tag{3.9}$$

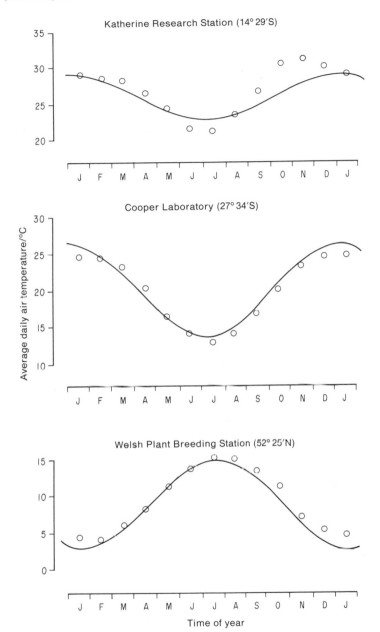

Fig. 3.7. Long-term averages of the monthly means of the average daily air temperature ($\bar{T}$) for three sites at different latitudes. The solid lines are obtained by applying eqns (3.7), (3.8) and (3.9) to the data for each site.

and these are illustrated in Fig. 3.8. Again, it needs to be emphasized that eqns (3.7)–(3.9) are only simple approximations. In the absence of direct measurements, they do no more than provide reasonable estimates of $\bar{T}$ at different times of year at different locations.

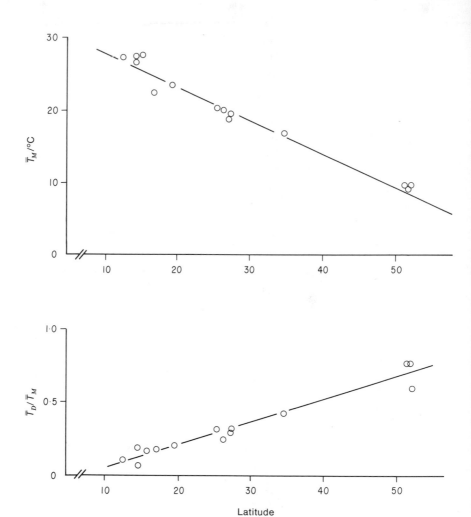

Fig. 3.8. Changes in the annual mean value for the average daily air temperature ($\bar{T}_M$) and the relative seasonal amplitude in the average temperature ($\bar{T}_D/\bar{T}_M$) with latitude. Solid lines are given by eqns (3.8) and (3.9).

3.4 Light interception by a "closed" crop

Synopsis: The proportion of the incident light flux density upon the surface of a "closed" crop canopy that is absorbed by it at any moment in time, Q, can be written as

$$Q = [1 - \exp(-kL)]$$

where k is a canopy light extinction coefficient and L is the leaf area index of the crop. The average proportion of the total light flux density incident upon the crop over the period of time t days, $\bar{Q}$, can be calculated as

$$\bar{Q} = 1 + [Q(0) - Q(t)]/\ln\{1 - Q(0)/[1 - Q(t)]\}$$

where $Q(0)$ and $Q(t)$ are the proportions of light intercepted by the crop at the beginning and end of the period. If, during the vegetative growth period, the crop attains its maximum standing leaf area at some time z days, $\bar{Q}$ may be better calculated as

$$\bar{Q} = [z\bar{Q}(z) + (t - z)Q(z)]/t$$

where $\bar{Q}(z)$ is the value of $\bar{Q}$ obtained over the time interval $0-z$ days.

The downward light flux density within a "closed" crop canopy beneath a cummulative leaf area index L, I, is usefully described by the simple equation

$$I = I_0 \exp(-kL) \tag{3.10}$$

where I_0 is the downward light flux density incident at the top of the canopy and k is a canopy light extinction coefficient. Saeki (1963) has proposed that the light flux density incident upon a leaf surface and available for photosynthesis, I_p, can be written as

$$I_p = [(1 + r)kI_0 \exp(-kL)]/(1 - m) \tag{3.11}$$

where r is a leaf reflectance coefficient and m is a leaf transmission coefficient. The derivation of eqn (3.11) from eqn (3.10) is discussed in more detail by Thornley (1976, Ch. 3).

Implicit in these equations is the assumption that the downward light flux density in any horizon within the "closed" crop canopy is spatially homogeneous. Kasanaga and Monsi (1954) and Monteith (1965) have looked at the problem of "sunflecking" within the canopy. The downward light flux density falling on the nth plane within a canopy diminishes exponentially with the number n according to

$$I = I_0[s + (1 - s)m]^{n-1} \tag{3.12}$$

where $(1 - s)$ is the fraction of the light intercepted by each unit leaf layer, essentially the ratio of the leaf area in any particular plane to the area of ground beneath the plane. We can write

$$\exp(A) = [s + (1 - s)m]^{n-1} \tag{3.13}$$

(so that eqn (3.12) becomes $I = I_0 \exp(A)$) and we can re-write eqn (3.13) as

$$A = (n - 1)\ln[s + (1 - s)m]. \tag{3.14}$$

Now the cummulative leaf area index above the nth plane, L, will be given by the simple product of $(n - 1)$ and $(1 - s)$, that is

$$L = (n - 1)(1 - s) \tag{3.15}$$

so that eqn (3.12) can be re-written as

$$\begin{aligned} I &= I_0 \exp(A) \\ &= I_0 \exp\{[L/(1 - s)]\ln[s + (1 - s)m]\} \end{aligned} \tag{3.16}$$

where we have used eqn (3.15) to replace $(n - 1)$ in eqn (3.14) by $L/(1 - s)$. If we now compare eqn (3.16) with eqn (3.10) we reveal the simple identity

$$k = [1/(1 - s)]\ln[s + (1 - s)m]. \tag{3.17}$$

The important point of this exercise is that eqns (3.10) and (3.12) are shown to be formally identical, and the canopy light extinction coefficient can be related to the amount of sunflecking within the canopy.

The proportion of the incident light flux density incident upon, and intercepted by, a "closed" crop at any instant in time, Q, can now be calculated. We can write that

$$Q = 1 - (I/I_0) = 1 - \exp(-kL). \tag{3.18}$$

If Q is measured on several occasions during the growth of a particular crop, we can use numerical techniques to estimate the average proportion of the light energy incident upon the crop that is intercepted by it during growth, and we can denote this proportion by $\bar{Q}$. For example, if we have n measurements of Q at equal time intervals, h, during crop growth, we can write that

$$\bar{Q} = \left[2 \sum_{i=1}^{i=n} Q_i - (Q_1 + Q_n) \right] \bigg/ 2(n-1). \tag{3.19}$$

Naturally, the greater n, that is the more frequent the measurements of Q, the more accurately we can expect to estimate $\bar{Q}$. However, we can use eqn (3.10) to provide a reasonable estimate of $\bar{Q}$ from more fragmentary data.

Most agronomists recognize that standing leaf dry weight of a "closed" crop (expressed as weight per unit of ground area) increases approximately linearly with time to some maximum value and then remains constant at this value for an extended period of time. We can describe such growth with the two equations

$$W_L = W_{L0} + mt \qquad 0 < t < z \tag{3.20a}$$

$$W_L = (W_L)_{\max} \qquad t > z \tag{3.20b}$$

where W_L is the leaf dry weight per unit ground area at time t, W_{L0} is the initial leaf weight (at time $t = 0$) and $(W_L)_{\max}$ is the maximum standing leaf weight realized at some time $t = z$. If we now assume that the specific leaf area of the crop, s_A, and the canopy light extinction coefficient, k, both remain constant throughout growth, we can use eqns (3.10) and (3.20) to write the proportion of the incident light flux density absorbed at any time t, $Q(t)$, as

$$Q(t) = 1 - \exp\left[-ks_A(W_{L0} + mt)\right] \qquad 0 < t < z \tag{3.21a}$$

$$Q(t) = 1 - \exp\left[-ks_A(W_L)_{\max}\right] \qquad t > z \tag{3.21b}$$

where we have replaced L in eqn (3.18) by the two relationships $L = s_A W_L$ and $L_{\max} = s_n(W_L)_{\max}$. We can also write that

$$\bar{Q} = \left[\int_0^t Q(t)\, dt \right] \bigg/ t \tag{3.22}$$

and for values of $t < z$, if we use eqn (3.21a) to substitute for $Q(t)$ in eqn (3.22) and then integrate, we obtain the expression for $\bar{Q}$,

$$\bar{Q} = 1 + [Q(0) - Q(t)]/\ln\{[1 - Q(0)]/[1 - Q(t)]\} \tag{3.23}$$

where $Q(0)$ and $Q(t)$ are measured values of Q at times $t = 0$ and $t = t$. For values of $t > z$, that is when the crop has grown for a sufficient length of time to attain its maximum standing leaf weight, we can use eqns (3.21a), (3.22) and (3.23) to obtain the expression

$$\bar{Q} = [z\bar{Q}(z) + (t - z)Q(z)]/t \tag{3.24}$$

where $\bar{Q}(z)$ is the value of $\bar{Q}$ obtained from eqn (3.23) when $t = z$.

The practical application of eqns (3.20) and (3.21) to experimental data for a vegetative mungbean crop are illustrated in Fig. 3.9. $\bar{Q}$ calculated over the interval 19–53 days was equal to 0.71 and over the whole vegetative growth period, 19–60 days, it was equal to 0.75.

The measurement of the proportion of incident light intercepted by a crop requires a certain amount of sophisticated equipment. If the measurements are combined with the measurement of the crop's leaf area index, to

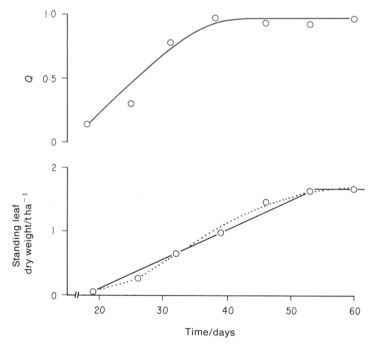

Fig. 3.9. Temporal changes in the standing leaf dry weight and the proportion of the incident light flux density intercepted (Q) for a mungbean crop. The solid lines are given by eqns (3.20) and (3.21).

then calculate the canopy light extinction coefficient, the measurements are demanding in time and labour. If there are no means available of measuring $Q(t)$, we may still be able to estimate $\bar{Q}$, the average proportion of the incident light intercepted during the vegetative growth phase. The estimate is based on three assumptions:

(i) that at emergence we can assume that $Q(0) = 0$;
(ii) that at the end of the vegetative growth phase, or when the crop has reached its maximum standing leaf weight (or leaf area) we can make a reasonable visual estimate of the proportion of incident light intercepted by it;
(iii) that we can reasonably accurately assess the time at which the crop attains its maximum standing leaf weight (or leaf area).

The estimate provides a pragmatic solution in the event that time or resources prevent the physical measurement of $Q(t)$. It is based on only two value judgements, $Q(t)$ at the end of vegetative growth or $Q(z)$, and the time z. Both these judgements may be made with reasonable accuracy by experienced agriculturists.

3.5   Light interception by a "spaced" plant

Synopsis: All crops are initially open arrays of plants. The rate of establishment of a "closed" crop surface, intercepting the greatest proportion of the incident light energy, is an important effector of the crop growth rate. The specific leaf area (leaf area per unit leaf dry weight) is an index of the plant's investment in developing the maximum light-intercepting surface area. It is markedly affected by the light and temperature regimens under which the plant is growing.

All crops are essentially open arrays of spaced plants during the early stages of their growth, each of the component plants intercepting light more or less independently of one another. The proportion of the incident light flux density per unit of ground area intercepted by any plant will be proportional to its leaf area. Thus we can write that

$$Q = bA \qquad (3.25)$$

where A is the total leaf area of the plant and b is a constant. Essentially b is the ratio of the leaf area of the plant projected on to a horizontal surface beneath it to its actual leaf area (see Section 2.1). It is useful to note here that when the product of the canopy light extinction coefficient and the leaf area index, kL, is small, eqn (3.10) reduces to the approximate form

$$Q = kL \qquad (3.26)$$

which probably provides a good approximation to eqn (3.25).

The daily increment in plant leaf area, ΔA, for an increment in standing leaf weight, ΔW_L, or shoot weight, ΔW_S, can be written as

$$\Delta A = s_A \Delta W_L = F_A \Delta W_S \qquad (3.27)$$

where s_A is the specific leaf area and F_A is the leaf-area ratio of the plant. The specific leaf area or leaf-area ratio of the plant therefore provides a simple index of its investment in developing leaf area for a given increment in leaf dry matter. In a teleological sense, a high specific leaf area would suggest that the plant is maximizing its light-intercepting, photosynthetic surface.

Both s_A and F_A are affected by the plant's aerial environment. These effects can be empirically described by an equation of the sort

$$s_A \text{ (or } F_A) = \bar{T}/(a_0 + a_1 \bar{T} + a_2 \bar{T}\bar{S}) \qquad (3.28)$$

where a_0, a_1, a_2 are constants and $\bar{T}$ and $\bar{S}$ are the average ambient temperature and daily light integral experienced by the developing leaves in the crop leaf canopy. Multiple regression equations of this type must be used with care when predicting values of the dependent variables s_A or F_A (Chanter, 1981). However, eqn (3.28) accounts for between 70% and 80% of the observed variation in s_A (or F_A) for a number of very different crops grown under a range of contrasting light and temperature regimens (Charles-Edwards, unpublished data). Estimates of the parameters a_0, a_1 and a_2 for tomato and chrysanthemum plants are shown in Table 3.2. These were obtained by the multiple regression of $(1/s_A)$ on $(1/\bar{T})$ and $\bar{S}$.

Table 3.2

Estimates of the coefficients a_0, a_1 and a_2 obtained by the multiple regression of $(1/s_A)$ on $(1/\bar{T})$ and $\bar{S}$, where $\bar{T}$ and $\bar{S}$ are the mean temperature and mean daily light integrals experienced by the plants during growth.

Plant	$a_0/10^3 \text{ g m}^{-2} (^\circ C)$	$a_1/10^3 \text{ g m}^{-2}$	$a_2/10^3 \text{ g MJ}^{-1}$
Tomato	0.50	-9.7×10^{-3}	10.0×10^{-3}
Chrysanthemum	0.46	-8.5×10^{-3}	8.4×10^{-3}

Genetic or environmentally induced differences in s_A (or F_A) will affect the amount of leaf area subtended by a plant for a given increment in leaf (or shoot) dry weight. Differences in s_A or F_A will clearly affect seedling growth and development, and thence the establishment of the crop's photosynthetic surface. For example, eqn (3.28) suggests that in general s_A decreases with lowering ambient temperatures. The leaf area subtended for a given increment in leaf dry weight, and thence the proportion of incident light intercepted by the plant, will therefore most probably decrease with decreasing temperatures, and we might expect the establishment of a crop to be thereby adversely affected. Whereas the rate of light-saturated leaf photosynthesis is often highly and positively correlated with s_A, selection of genotypes with this character in an attempt to increase the plant's photosynthetic performance might not be reflected in improved plant growth rates because of concomitant effects on the rate of leaf-area development by the plant (Charles-Edwards, 1979). This latter point has been discussed more fully in Section 2.1.

3.6 Light interception by isolated stands of plants

Synopsis: Since many crops are grown commercially as discrete blocks or rows of plants we need to develop methods of estimating the proportion of the incident radiant energy intercepted by them. Whereas good estimates of the patterns of light attenuation within a crop canopy can be obtained if the geometry of the canopy can be described, complex computer programs, involving long numerical integrations, are needed to make these estimates. Although simple to make in principle, the estimates are difficult and complex to make in practice. For the routine analysis of field data we need simpler, more pragmatic methods of calculating the proportion of incident light intercepted by a row crop.

Many crops are grown commercially as blocks or rows of plants, and the penetration of light between adjacent blocks or rows can lead to considerable spatial heterogeneity in the downward light flux density within their leaf canopies. This spatial heterogeneity will clearly affect their photosynthetic performance, and we need some quantitative understanding and treatment of this effect.

Let us suppose that the leaf canopy is made up of a large number of small leaves which are randomly distributed throughout its volume. In this case, if we denote $A(x,y,z)$ as the leaf area per unit volume of canopy

$$A(x,y,z) = A = \text{constant.} \tag{3.29}$$

If the light intensities (W m^{-2} steradian^{-1}) in the direction $P \rightarrow Q$ are I_P^*

and I_Q^* at the points P and Q respectively, we can write

$$I_Q^* = I_P^* \exp(-kL), \tag{3.30}$$

where L is the leaf area index along the path PQ, projected onto a plane normal to the path, and k is a light extinction coefficient. The leaf area index, L^*, transversed along the path of length s within the leaf canopy, can be obtained by integrating the leaf-area density function, $A(x,y,z)$ along the path, that is

$$L^* = \int_0^s A(x,y,z)\,ds \tag{3.31}$$

and if we use eqn (3.29) to define $A(x,y,z)$

$$L^* = As. \tag{3.32}$$

If the leaves are randomly inclined and oriented, the projected leaf area index L, normal to the path PQ, is given by

$$L = L^*/2 = As/2, \tag{3.33}$$

and we can substitute for L in eqn (3.30) to obtain

$$I_Q^* = I_P^* \exp(-kAs/2). \tag{3.34}$$

Now the downward light flux density upon an unobstructed horizontal plane, I_0, is given by

$$I_0 = \int_0^{2\pi} \int_0^{\pi/2} B \cos\theta \sin\theta\, d\theta\, d\phi \tag{3.35}$$

where $B(\theta, \phi)$ is a function describing the luminosity of the sky in the direction specified by (θ, ϕ). The angle θ is the zenithal angle of a point in the sky, that is its angle from the vertical, and ϕ is its azimuthal angle, that is the angle in the horizontal plane from some specified direction (say North or South), when observed from a particular point (x,y,z) within the canopy. For the obstructed point on an horizontal plane, the point Q, eqn (3.35) becomes

$$I_Q = \int_0^{2\pi} \int_0^{\pi/2} B \exp(-kAs/2) \cos\theta \sin\theta\, d\theta\, d\phi. \tag{3.36}$$

Although it is mathematically daunting, eqn (3.36) simply represents the contributions to the downward light flux density at Q of the separate intensities from all directions in the solid angle (θ, ϕ). One of the main problems for the analyst is to define how the path-length s depends upon the angles θ and ϕ; that is how it depends upon the coordinates of the point Q and the shape of the canopy.

For a well-defined canopy shape, such as a rectangular block of plants or an hedgerow orchard this problem is one of coordinate geometry and algebra (cf. Acock *et al.*, 1978; Charles-Edwards and Thorpe, 1976). The calculations are complicated, but the mathematics is quite straightforward. Simulations and measurements of the light-transmission profiles beneath a rectangular block and double-row of chrysanthemum plants is illustrated in Fig. 3.10. The "skewness" of the measured profile is almost certainly attributable to a strong direct beam component of sunlight that was not accounted for in the simulation. The canopies of the trees in the hedgerow apple orchard were described geometrically as a set of parallel, long cylinders, with ellipsoidal cross-section. Transmission profiles were measured within the orchard on a cloudy day, when there was no direct sunlight, and on two occasions when there was a strong direct beam component of sunlight. The transmission profiles simulated for these three occasions are compared with the measured profiles in Fig. 3.11. It can be seen that there is good agreement between the simulated and predicted profiles.

Another problem is to define B, the luminosity of the sky. For a standard overcast sky, B is usually taken to be a function of the zenithal angle, θ, alone, such that

$$B = B_0(1 + 2\cos\theta)/3. \tag{3.37}$$

Equation (3.37) mathematically represents the assumption that the luminosity of the sky at the zenith ($\theta = 0$) is three times its luminosity at the horizon ($\theta = \pi/2$). Substitution of eqn (3.37) into eqn (3.35), followed by integration over the solid hemisphere, gives the downward light flux density upon an unobstructed horizontal surface as

$$I_0 = 7\pi B_0/9. \tag{3.38}$$

There may also be an additional component at the point Q due to direct sunlight. If I_s is the light flux density on an unobstructed plane perpendicular to the sun's direction, θ_s, this component, which is simply added to I_Q, will be given by

$$I_s = \cos\theta_s \exp(-kAs/2). \tag{3.39}$$

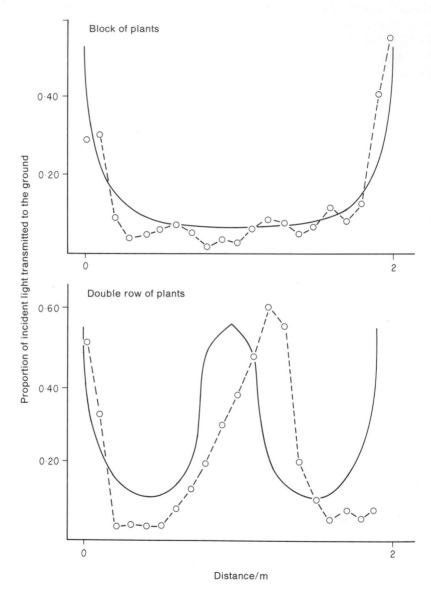

Fig. 3.10. Light transmission at the bases of an isolated block and double row crop of chrysanthemum plants. Measured values (○); simulated transmission profiles (——). The "skewness" of the measured transects could be attributed to a directional component of the incident light not taken into account in the simulations.

The development of these analyses and their application are dealt with in greater detail in Charles-Edwards (1981, Ch. 3).

For the routine analysis of crop field data we need a simpler, more pragmatic approach to estimating the proportion of the incident light energy intercepted by row crops. Jackson and Palmer (1979, 1981) have described a simple approach that can be used usefully in this context, and I will develop it now. They assume that the light transmitted through a discontinuous canopy, such as a row crop, Tr, has two separate and additive components. Firstly, there is the component which has passed between the rows of plants and is transmitted to the ground, even if the rows of plants absorb all the light that strikes them. We can denote this component by T_f. Secondly, there is the component that has been transmitted through the crop foliage, T_c. We can then write that

$$Tr = T_f + T_c. \tag{3.40}$$

The problem now is to relate T_c to the area and leaf arrangement of the crop canopy. They have proposed that as a good first approximation

$$T_c = (1 - T_f)\exp\left[-kL/(1 - T_f)\right] \tag{3.41}$$

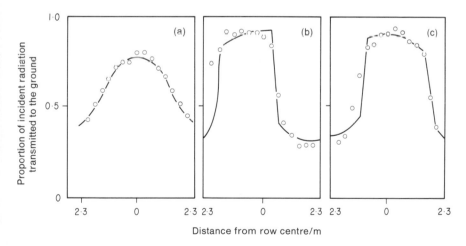

Fig. 3.11. Simulated (———) and measured (○) transmission profiles of radiation incident upon the ground beneath a hedgerow apple orchard, (a) under a diffuse sky condition and (b,c) on two occasions when there was a strong direct beam component of sunlight.

where k is a light extinction coefficient for the crop and L is its leaf area index. Combining eqns (3.40) and (3.41) yields

$$Tr = T_f + (1 - T_f)\exp[-kL/(1 - T_f)]. \tag{3.42}$$

A comparison of observed values of Tr, and values calculated using eqn (3.42) for hedgerow apple orchards can be made from their data and is shown in Fig. 3.12. Whereas the changing leaf area index of the crop can be calculated in the way described in Section 2.3, changes in T_f will depend upon the changing height and row widths as the crop develops.

For a uniform overcast sky, that is one for which the luminosity is constant across all zenithal and azimuthal angles, T_f can be written as the proportion of the ground area covered by the crop to the total ground area. For a standard overcast sky, or when there is a direct beam component of sunlight, we cannot necessarily make this simple calculation. For example, if the sun's zenithal angle is θ_s, direct sunlight will only be transmitted directly to the ground if

$$\theta_s < \tan^{-1}(w/b), \tag{3.43}$$

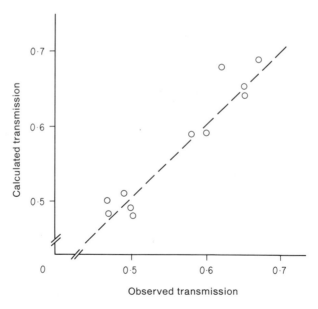

Fig. 3.12. A comparison of the observed and calculated (using eqn (3.42)) proportions of the incident light flux density transmitted to the ground (Tr) by a hedgerow apple orchard (after Jackson and Palmer, 1981).

where b is the height of the crop and w is the width of the gap between adjacent rows of plants.

For studies of rapidly growing crops it is probably an adequate approximation to assume that T_f is directly related to the proportion of the ground covered by the crop.

4

Light-Utilization Efficiency

4.1 Instantaneous rates of crop photosynthesis

Synopsis: The instantaneous rate of net photosynthesis per unit of ground area by a "closed" crop canopy can be written as a function of leaf photosynthetic characters, canopy architecture, the light flux density that is incident upon the crop and the proportion of the incident light flux density that is intercepted by it. The relationship between the net rate of canopy photosynthesis and the incident light flux density can be readily linearized, enabling the simple analysis of crop photosynthesis data.

A distinctive property of plants, and of all other living organisms, is that they are ordered and capable of creating order from their less-ordered surroundings. This property of orderliness reflects the fact that they are thermodynamically open systems and depend upon some external source of energy for their functioning. This external energy source is the sun, and the photosynthetic process is almost unique as a means by which the sun's transient, radiant energy is converted to a stable, storable, biologically accessible form, available for use by living organisms. All living organisms gain access to the sun's energy by consuming, either directly or indirectly, plants or plant parts. It is appropriate that any analysis of plant or crop

65

growth should be based upon an examination of the efficiency with which the incident light energy is used by the crop in the production of new plant dry matter. The conversion of the incident light energy to available chemical energy is effected by the process of photosynthesis.

It has been shown by Acock and co-workers that a number of empirical and mechanistic models of canopy photosynthesis data can provide equally good quantitative descriptions of the instantaneous response of canopy net photosynthetic rates to changes in the incident light flux density (Acock *et al.*, 1976). Mechanistic models of canopy net photosynthesis, based on the single leaf's photosynthetic behaviour, can, if properly formulated, provide consistent estimates of the leaf photosynthetic parameters when used to analyse canopy photosynthesis data (see Charles-Edwards, 1981, Ch. 3). It has been recently demonstrated that the instantaneous response of the rate of canopy photosynthesis by a "closed" crop of tomato plants to changes in the light flux density incident upon the crop can be successfully predicted from the photosynthetic behaviour of single, unshaded tomato leaves by the equation

$$F_C = \alpha I_0 F_{max}[1 - \exp(-kL)]/(\alpha k I_0 + F_{max}) - R_C \qquad (4.1)$$

where F_C and R_C are the rates of canopy net photosynthesis and dark respiration per unit of ground area, k is a canopy light extinction coefficient, L is the leaf area index of the crop, I_0 is the downward light flux density incident at the top of the crop, α is the leaf photochemical efficiency and F_{max} is the rate of light-saturated photosynthesis per unit leaf area by an upper, unshaded leaf in the crop canopy (see Charles-Edwards, 1981, Ch. 3, Section 4). Equation (4.1) is derived from the single-leaf light-response function,

$$F = \alpha I F_{max}/(\alpha I + F_{max}) - R \qquad (4.2)$$

the classical hyperbolic response function of leaf photosynthesis, F, to changing incident light flux density, I. Equation (4.1) is then derived from eqn (4.2) on the assumption that the rate of light-saturated photosynthesis of any leaf within the canopy is directly proportional to that proportion of the light flux density incident upon the crops uppermost surface that is transmitted to the leaf *in situ* within the canopy. The assumption is supported by the data shown in Table 4.1. Estimates of the rates of light-saturated photosynthesis for tomato leaves from different positions within the same stand of plants, together with their leaf photochemical efficiencies, are shown in the Table. The marked variation in leaf photosynthesis appears to be associated with leaf position rather than with leaf age. An old leaf (the ninth leaf produced) from the outside edge of a guard plant had a higher rate of light-saturated photosynthesis than a young leaf (the twentieth leaf produced) from a shaded position within the crop canopy. In contrast, other old leaves (numbers 12 and 9) from heavily shaded positions toward the base of the canopy had lower rates of light-saturated photosynthesis than the old leaf from the guard plant.

Equation (4.1) has the important virtue of simplicity. It can readily be transformed into a linear relationship between F_C and I_0 by taking reciprocals of both sides, when

$$[1 - \exp(-kL)]/(F_C + R_C) = k/F_{max} + 1/\alpha I_0. \tag{4.3}$$

If we replace the term $[1 - \exp(-kL)]$ by Q, the proportion of the incident downward light flux density intercepted by the crop, eqn (4.3) can be re-written in the simpler form

$$Q/(F_C + R_C) = k/F_{max} + 1/\alpha I_0. \tag{4.4}$$

The proportion of the incident light energy intercepted by a crop, Q, can be quite simply measured directly as the difference between the downward light flux density incident upon the crop and the sum of the upward light flux density reflected off the canopy surface and the downward light flux density transmitted to the soil surface beneath the crop. Essentially it requires three simultaneous "light measurements", and is experimentally less tedious and time consuming to obtain than crop leaf area indices and light extinction coefficients.

Table 4.1

Estimates of photochemical efficiency (α) and the rate of light-saturated leaf photosynthesis (F_{max}) for leaves of different ages and from different horizons within a stand of tomato plants.

Leaf number	$\alpha/\mu g\,(CO_2)\,J^{-1}$	$F_{max}/10^{-3}\,g\,(CO_2)\,m^{-2}\,s^{-1}$	Note
20	11.4 (± 1.2)	1.08	(a)
20	13.1 (± 2.9)	0.36	
12	12.2 (± 3.0)	0.24	(b)
9	10.0 (± 4.0)	0.12	
9	15.6 (± 2.6)	0.66	(c)

(a) Upper unshaded leaf.
(b) Shaded leaf.
(c) Partially shaded leaf from outer guard plant.

The linear regression of $Q/(F_C + R_C)$ on $1/I_0$ enables canopy photosynthesis data obtained on "closed" crops in the field to be analysed in terms of the leaf photosynthetic characters (F_{max} and α) and the canopy architecture (k). For example, it has been used to analyse the main effects of the height and frequency of defoliation on dry-matter production by tropical grass/legume swards (Ludlow and Charles-Edwards, 1980). It has also been used to examine the effects of sward structure on light distribution within the canopy (k) and thence canopy photosynthesis. The application of eqn (4.4) to data for a "closed" crop of tomato plants is illustrated in Fig. 4.1. The slope of the regression line provides an estimate of $\alpha = 11.5\ \mu g(CO_2)\ J^{-1}$ for tomato leaves, a value which corresponds well with estimates of α obtained by the analysis of single-leaf photosynthesis data for these plants. For example, the mean of the values of α shown in Table 4.1 is approximately $12.5\ \mu g(CO_2)\ J^{-1}$.

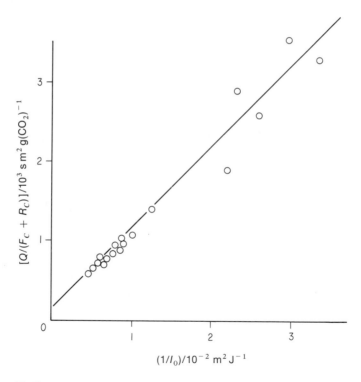

Fig. 4.1. The linear regression of $Q/(F_C + R_C)$ on $1/I_0$ for canopy photosynthesis data collected on a "closed" tomato crop.

4.2 The daily gross photosynthetic integral

Synopsis: If it is assumed that the diurnal variation in the downward light flux density incident upon the ground varies sinusoidally during the day, a simple algorithm for the daily net photosynthetic integral can be obtained. A suite of analogous equations describing the instantaneous rates of leaf, F, and canopy, F_C, net photosynthesis and the daily net photosynthetic integral, ∇_F, can be obtained as

$$F = \alpha I F_{max}/(\alpha I + F_{max}) - R$$

$$F_C = \alpha I_0 F_{max}[1 - \exp(-kL)]/(\alpha k I_0 + F_{max}) - R_C$$

$$\nabla_F = \alpha S F_{max} h[1 - \exp(-kL)]/(\alpha k S + h F_{max}) - \nabla_R$$

where α is the leaf photochemical efficiency, F_{max} the rate of light-saturated photosynthesis by an upper unshaded leaf in the canopy, k is a canopy light extinction coefficient, L is the crop's leaf area index, I and I_0 are the light flux densities incident upon the leaf and crop respectively, S is the daily light integral, h is the daylength, R and R_C are the instantaneous rates of leaf and crop respiration and ∇_R is the daily respiratory integral.

We need to be able to integrate eqn (4.1) over the course of the day to obtain an expression for the daily gross photosynthetic integral of a crop. The integration requires a knowledge of the diurnal variation in the downward light flux density incident upon the crop. Monteith (1965) has suggested that for cloudless days in all climates, and for average days in climates where cloud cover remains constant throughout the day, the diurnal variation in the downward light flux density on an unobstructed horizontal surface, I_0, can be described by the simple sine function

$$I_0 = [\pi S/2h] \sin(\pi t/h) \qquad 0 < t < h \qquad (4.5a)$$

$$I_0 = 0 \qquad h < t < 86{,}400 \qquad (4.5b)$$

where h is the daylength (in seconds), t is the time after first light and S is the daily integral of the downward light flux density incident upon an unobstructed surface at ground level.

We can use eqn (4.5a) to substitute for I_0 in eqn (4.1), when we can obtain the daily *gross* photosynthetic integral, ∇_C, as

$$V_C = \int_0^h (F_C + R_C)\, dt \qquad 0 < t < h,$$

$$= \{F_{max}[1 - \exp(-kL)]/k\} \int_0^h \{\sin(\pi t/h)/[\delta + \sin(\pi t/h)]\}\, dt \quad (4.6)$$

where $\delta = 2hF_{max}/k\pi\alpha S$. The problem is one of finding the integral of the sine function on the right-hand side of eqn (4.6). This is a standard integral, and the solution can be found in most tables of standard integrals. The daily gross photosynthetic integral can be obtained from eqn (4.6) as

$$V_C = 2h[(\pi/2) - \S]\, F_{max}[1 - \exp(-kL)]/\pi k \qquad (4.7)$$

where we can calculate solutions from

$$\S = [2\delta/\sqrt{(\delta^2 - 1)}]\, \text{arctg}\,[\sqrt{(\delta^2 - 1)/(\delta + 1)}] \qquad \text{when } \delta^2 > 1$$

$$\S = 1 \qquad \text{when } \delta^2 = 1$$

$$\S = \delta/[\sqrt{(1 - \delta^2)}]\, \ln\{[1 + \delta + \sqrt{(1 - \delta^2)}]/[1 + \delta - \sqrt{(1 - \delta^2)}]\}$$

$$\text{when } \delta^2 < 1.$$

The integral is fairly cumbersome, hard to visualize, and therefore not too helpful to the non-mathematician. However, a simple and robust algorithm for it over all values $0 < \delta < \infty$ is given by

$$[(\pi/2) - \S] = (\pi/2)/[1 + \delta(\pi/2)]. \qquad (4.8)$$

This algorithm can be compared with the exact integral in Fig. 4.2, and it can be seen to provide a reasonable, accurate and simple substitute for it over the whole range of values of δ.

We can substitute this algorithm for $(\pi/2 - \S)$ in eqn (4.7) to obtain an expression for the gross daily photosynthetic integral, V_C,

$$V_C = 2hF_{max}(\pi/2)[1 - \exp(-kL)]/k\pi[1 + \delta(\pi/2)]. \qquad (4.9)$$

We can now replace δ by $2hF_{max}/k\pi\alpha S$. If we do this, and rearrange the resulting expression, we can obtain

$$V_C = \alpha Sh F_{max}[1 - \exp(-kL)]/(k\alpha S + hF_{max}). \qquad (4.10a)$$

Now the daily integral for net photosynthesis, which we can denote by ∇_F, will be given by

$$\nabla_F = \nabla_C - \nabla_R \qquad (4.10b)$$

where ∇_R is the daily respiratory integral. We can then re-write eqn (4.10b) as

$$\nabla_F = \alpha ShF_{max}[1 - \exp(-kL)]/(k\alpha S + hF_{max}) - \nabla_R. \qquad (4.10c)$$

It is useful and instructive to compare eqns (4.2), (4.1) and (4.10c):

$$F = \alpha I F_{max}/(\alpha I + F_{max}) - R \qquad (4.2)$$

$$F_C = \alpha I_0 F_{max}[1 - \exp(-kL)]/(\alpha k I_0 + F_{max}) - R_C \qquad (4.1)$$

$$\nabla_F = \alpha ShF_{max}[1 - \exp(-kL)]/(\alpha kS + hF_{max}) - \nabla_R \qquad (4.10c)$$

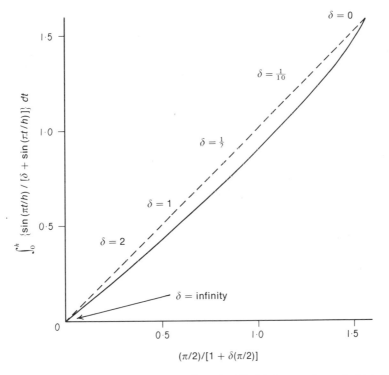

Fig. 4.2. A comparison of the algorithm $(\pi/2)/[1 + \delta(\pi/2)]$ with the exact integral of $\{\sin(\pi t/h)/[\delta + \sin(\pi t/h)]\}\,dt$ (given by $[(\pi/2) - \S]$ see eqns (4.6) and (4.7)) over the range of values of δ, $0 < \delta < \infty$.

There is a remarkable similarity in the form of the three equations. In going from eqn (4.2) to eqn (4.1) we have integrated photosynthesis over space. In going from eqn (4.1) to (4.10c) we have integrated over time. The similarity in the resulting equations is pleasing and intuitively satisfying.

The term $S[1 - \exp(-kL)]$ in eqn (4.10c) describes the amount of light energy incident upon the crop during the day which is absorbed by it. If we replace the term $[1 - \exp(-kL)]$ by Q, the proportion of the incident light energy absorbed, we can re-write eqn (4.10b) as

$$V_F = \alpha Sh F_{max} Q/(\alpha kS + h F_{max}) - V_R \tag{4.11}$$

We can now divide the top and bottom of the right-hand side of eqn (4.11) by h, the daylength, when we have

$$V_F = \alpha S F_{max} Q/(\alpha k \bar{I}_0 + F_{max}) - V_R \tag{4.12}$$

with the average incident light flux density, $\bar{I}_0$, equal to S/h. Equation (4.12) describes the daily *net* photosynthetic integral in terms of some well-defined leaf and canopy characters, α, F_{max} and k. It provides a simple and useful starting point for our mechanistic description of the crop light-utilization efficiency.

4.3 The daily respiratory integral

Synopsis: The McCree equation, which relates the rate (or amount) of crop respiration to both the standing crop biomass and the rate (or amount) of crop gross photosynthesis, may not be entirely appropriate for the calculation of daily respiratory integrals. A modified version of the McCree analysis may be more robust and useful for an analysis of crop dry-matter production.

McCree (1970) has related the respiratory losses of carbon dioxide by plants to both their photosynthetic activity and to their net dry weight by an equation of the form

$$V_R = aV_C + bW \tag{4.13}$$

where V_R denotes the daily integral of the respiratory loss, V_C is the daily gross photosynthetic integral, W is the plant dry weight and a and b are constants. Hansen and Jensen (1977) have applied eqn (4.13) to data for seedlings of *Lolium multiflorum* growing in different light regimens and have shown that it provides a good description of their data. However, the values of the coefficients *a* and *b* that they obtained after analysis of their data differed with changing daylength, one of the treatments in their experiments. They subjected plants to alternate three-day cycles of high and low light at a constant daylength, the daylength being changed between experiments. One set of their experimental observations is illustrated in Fig. 4.3. If all the data from their different experiments are pooled, the simple regression of V_R on V_C provides a reasonable description of them (see Fig. 4.4). If V_R is regressed on V_C separately for data obtained (a) on the first day of low light following a period of high light and (b) on the first day of high light following a period of low light, the two regressions provide reasonable error bounds for the regression of V_R on V_C for all the data. These separate regressions are illustrated in Fig. 4.5. On a dull day following a series of bright days, the respiratory losses are relatively high, and we can conceive that this arises because the plant utilizes stored assimilate for the processes of growth and maintenance of its metabolic machinery. In contrast, on a bright day following a series of dull days, the losses are relatively low. We can conceive that the plant is then using the newly assimilated material to replenish its stored reserves. This simple concept is tenable provided that the synthesis and storage of carbohydrate reserves are less energetically demanding upon the plant that its other metabolic functions.

McCree's analysis has led to the recognition of two components of plant/crop respiratory activity. One of these components, called the maintenance respiration component, is generally related to the net weight of the plant or crop. However, it does not necessarily follow that the size of the metabolically active component of a plant or crop is related to its total dry weight, and following Thornley's reinterpretation of plant respiration (Thornley, 1977), Barnes and Hole (1978) have suggested that eqn (4.13) may be more usefully written as

$$V_R = aV_C + bN \tag{4.14}$$

where N is the amount of proteinaceous material in the plant.

For many purposes it is probably adequate to approximate eqn (4.14) by the simple expression

$$V_R = aV_C + \text{constant.} \tag{4.15}$$

For example, measured values of V_R and V_C for a vegetative, "closed" crop of chrysanthemum plants growing under natural daylight give rise to the empirical relationship for this crop

$$V_R = 0.33V_C + 2.0 \tag{4.16}$$

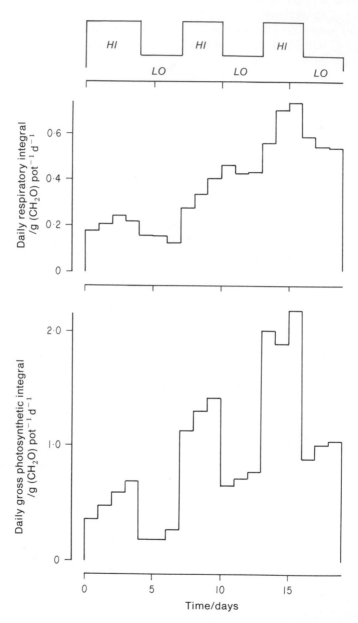

Fig. 4.3. Daily respiratory and gross photosynthetic integrals of seedlings of *Lolium multiflorum* grown under successive periods of three days at high light and three days at low light.

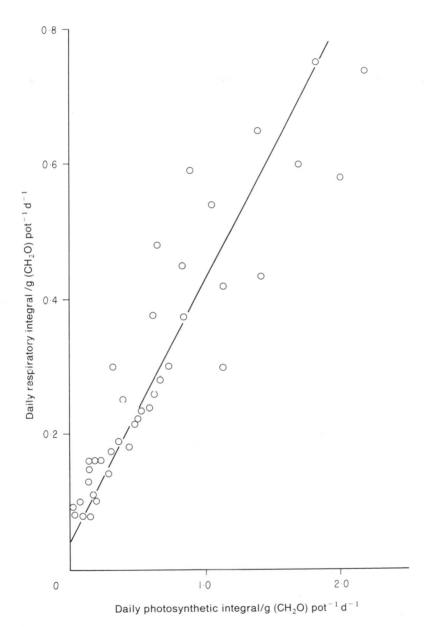

Fig. 4.4. The observed relationship between the daily respiratory integral and the gross photosynthetic integral of seedlings of *Lolium multiflorum* (for details see text).

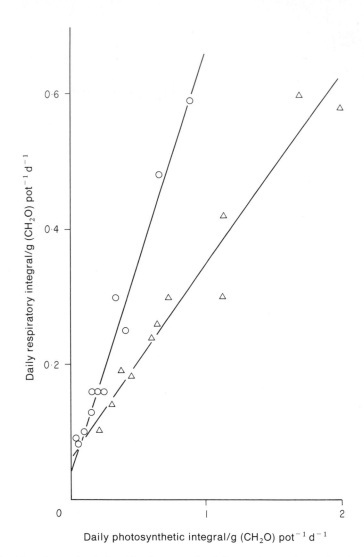

Fig. 4.5. The observed relationships between the daily respiratory integral and the gross photosynthetic integral on a dull day following three successive bright days (○), and on a bright day following three successive dull days (△).

where the constant (numerical value 2.0) has units g (CO_2) m^{-2}d^{-1} (see Acock *et al.*, 1979). It is a fairly common observation that the amount of nitrogen contained in the above-ground parts of a crop does not increase linearly with increasing crop dry weight per unit ground area, but reaches a maximum value and may then decline with time from that maximum value. For example, Watson and co-workers (Watson *et al.*, 1963) have shown that there is a dramatic decrease in the proportion of nitrogen in plant dry matter, with time, in two winter and two spring wheat varieties grown in the United Kingdom. This is presumably the result of the metabolically active component of the crop becoming an increasingly smaller proportion of the total crop biomass. It does not seem unreasonable to suppose that as a useful first approximation we can relate the size of the metabolically active component of the crop to the amount of light energy currently being absorbed by the crop, J. If we make this assumption we can re-write eqn (4.14) as

$$V_R = aV_C + bJ. \qquad (4.17)$$

The amount of light energy absorbed by a crop does not increase linearly with standing crop dry weight but approaches a maximum value with increasing leaf area index according to

$$J = S[1 - \exp(-KL)] = QS \qquad (4.18)$$

(see Section 3.4 for an example of how Q changes with leaf area index development), and eqn (4.18) may provide a good approximation to the changes in crop nitrogen content during growth.

Greenwood and co-workers examined changes in the protein content of plants and field crops during growth (Greenwood and Barnes, 1978; Greenwood *et al.*, 1978), relating protein content and maintenance respiration rates. In essence, they assumed that the maintenance respiration rate is the rate at which the carbohydrate, identified as glucose, is metabolized in maintaining protein per gram of protein per day. This enabled them to relate the protein content of the plant to the total weight of nitrogen-free organic matter. They concluded that, as the maintenance respiration rate is very dependent upon temperature, the change in protein content of the plant during growth will also depend upon the growth temperature of the plant. Plant-growth models based on these relationships provide good descriptions of real plant growth data. Their analyses are, in one sense, the corollary of the analysis proposed above, and they use growth data to estimate a maintenance respiration coefficient.

4.4   The crop light-utilization efficiency

Synopsis: The crop light-utilization efficiency, ε, can be written as

$$\varepsilon = q\nabla_F/J$$

where q converts the daily *net* photosynthetic integral to an increment in new plant dry matter, ∇_F is the daily *net* photosynthetic integral and J is the amount of light energy intercepted by the crop during the day. Using the analyses developed in the previous sections, we can write the light-utilization efficiency of a "closed" crop as a function of leaf and canopy photosynthetic characters. Environmental and physiological effectors of the crop light-utilization efficiency can be examined at the detailed, physiological level.

The crop light-utilization efficiency has been previously defined (for example see Section 2.1) as the ratio of the gross increment in crop dry matter to the amount of light energy absorbed by the crop. Formally we can write that the crop light-utilization efficiency, ε, is given by

$$\varepsilon = q\nabla_F/J \qquad (4.19)$$

where q converts g (CO_2) to g (dry matter), ∇_F is the daily *net* photosynthetic integral of the crop, and J is the amount of light energy intercepted by the crop during the day. We can use eqn (4.10b) to re-write eqn (4.19) as

$$\varepsilon = q(\nabla_C - \nabla_R)/J \qquad (4.20)$$

and if we use eqn (4.17) to substitute for ∇_R, eqn (4.20) becomes

$$\varepsilon = q(\nabla_C - a\nabla_C - bJ)/J$$
$$= q[(1 - a)\nabla_C - bJ]/J \qquad (4.20)$$

where ∇_C is the daily *gross* photosynthetic integral. We can replace $(1 - a)$ by Y_g the growth yield of the crop (cf. Hunt and Loomis, 1979) and use eqn (4.12) to substitute for ∇_C in eqn (4.20), when we obtain

$$\varepsilon = [qY_g\alpha F_{max}/(\alpha k\bar{I}_0 + F_{max})] - qb. \qquad (4.21)$$

Equation (4.21) now relates the efficiency with which a "closed" crop uses intercepted light energy in the production of new dry matter to six well-defined leaf and canopy characters, and to the independent environmental variable $\bar{I}_0$.

We can use eqn (4.21) to calculate ε for a "closed" vegetative chrysanthemum crop grown under natural daylight at different air temperatures, and at one temperature but under different mean daily incident light flux densities. These calculations, based on the leaf and canopy photosynthesis data reported by Charles-Edwards and Acock (1977), are given in Table 4.2. Whilst ε increased with increasing air temperatures, five-fold differences in the mean daily light flux density had little effect on ε. These estimates of ε from basic plant and crop physiological characters provide us with an idea of the overall dependence of ε on the light and temperature environment of the crop. Whilst there may be marked effects of the cropping environment on the physiological and environmental effectors of the crop light-utilization efficiency, these results indicate the *net* effects of these on ε. It is useful to have some feel for the interactions that lead to these *net* effects before we examine the physiological and environmental effectors of ε separately.

Table 4.2

Estimates of the crop light-utilization efficiencies of "closed" vegetative chrysanthemum crops (ε) grown under natural daylight at different controlled air temperatures (T) and experiencing different mean daily light flux densities ($\bar{I}_0$). The light-utilization efficiencies were calculated from leaf and canopy photosynthesis data.

$T/°C$	$\bar{I}_0/J\ m^{-2}\ s^{-1}$	$\varepsilon/\mu g$ (dry matter) J^{-1}
10	97	1.9
15	97	2.0
20	97	2.2
25	97	2.4
30	97	2.4
20	34	2.2
20	73	2.3
20	97	2.2
20	152	2.2
20	165	2.2

One of the most obvious physiological effectors of the crop light-utilization efficiency is the rate of light-saturated leaf photosynthesis, F_{max}. It is amenable to direct measure, and there is a considerable body of information on the changes in F_{max} associated with changes in the plant's growth regimen. For example, the rate of light-saturated photosynthesis per unit leaf area, F_{max}, usually increases as the mean daily light flux density experienced by the leaf during growth increases (cf. Charles-Edwards, 1979; Charles-Edwards *et al.*, 1974; Prioul and Bourdu, 1973). If the relationship between F_{max} and $\bar{I}_0$ can be approximated by the simple linear relationship

$$F_{max} = a_0 \bar{I}_0 \qquad (4.22)$$

where a_0 is a proportionality constant, eqn (4.21) can be simplified to

$$\varepsilon = q Y_g \alpha a_0 / (\alpha k + a_0) - qb, \qquad (4.23)$$

and the explicit dependence of ε on $\bar{I}_0$ is removed (see the data in Table 4.2 which show little effect of $\bar{I}_0$ upon ε). Provided q, Y_g, α, k and b are not markedly dependent upon $\bar{I}_0$, there will be no implicit dependence of ε on $\bar{I}_0$ either. Whereas measured values of q for the chrysanthemum crops declined with increasing values of $\bar{I}_0$ there were no consistent changes in k (Acock *et al.*, 1979). These data are shown in Table 4.3. Recent studies of the seasonal changes in F_{max} of leaves from field-grown grass swards suggest that these changes may be associated with seasonal changes in $\bar{I}_0$ (Tateno and Iida, 1980).

The main effect of different air temperatures during growth on the leaf and canopy characters of the chrysanthemum plants appeared to be on q and k (see Table 4.3). As the air temperature increased, q increased and k decreased. The change in q must reflect a change in the metabolic balance of the plants with different growth temperatures. The conversion factor q is generally given a value of 0.68, the ratio of the molecular weights of the carbohydrate moiety (CH_2O) and carbon dioxide (CO_2). The data for the chrysanthemum suggest that this value may be too low, and a value of 0.71 or 0.72 might be more appropriate.

Equation (4.21) allows us to identify some basic, physiological effectors of the crop light-utilization efficiency, and we could use this equation to integrate our understanding of the more basic plant physiological processes into an understanding of the field performance of crops. It has been intuitively felt by many agriculturists that crop productivity could be increased by selecting plant genotypes with increased rates of leaf photosynthesis, and this feeling appears to be formalized by eqn (4.21). However, as has been discussed in Section 2.1, there is evidence that a considerable part of the genetic variation in F_{max} between cultivars may be associated with differences in their leaf thickness, the thicker leaved cultivars having higher values for F_{max} (Charles-Edwards, 1979) than the thinner ones. Thicker leaves imply lower rates of leaf-area expansion, and a longer time-period for a crop of thick-leaved cultivars to attain "full ground cover". An improvement in ε by selection for genotypes with a high F_{max} may therefore be offset by a reduction in the amount of light energy intercepted by crops of these genotypes during the early stages of growth and development.

Table 4.3

Measured values of the conversion constant (q) (g (dry matter) g $(CO_2)^{-1}$ assimilated) and canopy light extinction coefficient (k) for "closed" vegetative chrysanthemum crops grown under natural daylight at different controlled air temperatures (T) and under different mean daily light flux densities ($\bar{I}_0$).

$T/°C$	$\bar{I}_0/J\ m^{-2}\ s^{-1}$	q	k
10	97	0.70	0.71
15	97	0.72	0.66
20	97	0.73	0.53
25	97	0.75	0.49
30	97	0.76	0.46
20	34	0.77	0.57
20	73	0.73	0.58
20	97	0.73	0.53
20	152	0.71	0.49
20	165	—	0.56

F_{max} may also depend upon the nitrogen content of the leaf (see Charles-Edwards, 1981, Ch. 2; Gulson and Chu, 1981, Fig. 2). Leaf nitrogen content will depend upon the amount of nitrogen available to the plant and will also be subject to genetic variation between cultivars of the same crop. Soil fertility might then be expected to have a direct effect on the light-use efficiency of crops, but the effects may differ between cultivars of the same species. For example, a study has been made of a number of *Eucalypt* species originating from different habitats (Mooney *et al.*, 1978). Although the species were grown under the same experimental environment, the rates of light-saturated leaf photosynthesis, F_{max}, were not correlated with differences in specific leaf weight. However, the rate of light-saturated photosynthesis per unit leaf dry weight (the product $s_A F_{max}$) was highly correlated with the leaf nitrogen content. In general, the species with lower leaf nitrogen contents has lower specific leaf areas (thicker leaves). More crudely, the plants with lower specific leaf photosynthetic activities had larger photosynthetic systems beneath unit leaf area. This observation has obvious implications for this type of analysis of crop growth. Equation (4.21) provides us with a way of identifying and quantifying these effects on the light-utilization efficiency of the crop. From the practical, agricultural point of view, the measurement of the proportion of elemental nitrogen in the leaf and the specific leaf area of fully expanded leaves at the top of the canopy may provide useful information on factors affecting F_{max}, and thence ε.

Equation (4.21) also demonstrates how the distribution of light within the crop canopy can affect the crop light-utilization efficiency (see Section 2.3). For crops consisting of prostrate plant types, the canopy light extinction will be high. Since it appears in the denominator of eqn (4.21) a higher value for k will tend to reduce ε. For crops consisting of erect leaves the converse is true. For a low value of k, or a canopy of erect leaves, light is distributed more evenly throughout the canopy. But note that the effect of canopy architecture will depend upon the relative values of $\alpha \bar{I}_0 k$ and F_{max}. If $\alpha \bar{I}_0 k \ll F_{max}$, k will have little effect, since eqn (4.21) will reduce to

$$\varepsilon = q(Y_g \alpha - b). \tag{4.24}$$

If $\alpha \bar{I}_0 k \gg F_{max}$, however, the effects of k will be large since eqn (4.21) will reduce to

$$\varepsilon = q Y_g F_{max} / k \bar{I}_0 - qb. \tag{4.25}$$

4.5   Estimates of crop and plant light-utilization efficiencies from field data

Synopsis: Estimates of the crop light-utilization efficiency, obtained by calculation from basic physiological data, can be compared with those calculated directly from dry-matter data obtained for "closed" crops during growth. Higher estimates of ε, derived from a variety of published data, cannot be unambiguously accounted for.

In the previous section, we have looked at the physiological determinants of the crop light-utilization efficiency. It is always an uncertain step when we attempt to integrate our understanding of plant or organ processes up through several levels of organization into an understanding of the field behaviour of the plant or crop. However, we can directly estimate the light-utilization efficiency of a crop from field data. Efficiencies of light use in the production of above-ground parts by C_3 crops, obtained by analysis of field data, can be compared with those obtained for the chrysanthemum crops by calculation from fundamental physiological parameters. The estimates of ε, given in Table 4.2, can be scaled by the published values for the shoot partition coefficient (Acock *et al.*, 1979) to provide estimates of ε_T for these chrysanthemum crops. The estimate of ε obtained for a crop growth at a comparable temperature can then be compared with those obtained by analysis of field growth data for mungbean and *Stylosanthes humilis* (given in Table 2.8 in Chapter 2). This is done in Table 4.4 The estimates of ε_T for the four crop species are all very similar and suggest that eqn (4.21) has successfully integrated our knowledge of plant processes into a real understanding of an important physiological determinant of crop behaviour in the field.

Table 4.4

The light-utilization efficiencies of above-ground dry-matter production (ε_T) by "closed" vegetative chrysanthemum crops, mungbean (*V. radiata* and *V. mungo*) and *Stylosanthes humilis* at different controlled air temperatures (T) and under different mean daily light flux densities ($\bar{I}_0$)

$T/^\circ C$	$\bar{I}_0/\mathrm{J\ m^{-2}\ s^{-1}}$	$\varepsilon_T/\mu g$ (dry matter) $\mathrm{J^{-1}}$	Species
30	97	2.0	*C. morifolium*
29	195	2.2	*V. radiata*
29	214	2.2	*V. mungo*
28	200	1.7	*S. humilis*

Estimates of the efficiency of light use in the *net* production of crop dry matter, ε^*, can also be obtained quite simply from the regression of standing crop dry weight on the cumulated intercepted radiation. Allen and Scott (1980) have reported growth rates of potato crops at 20-day intervals between twenty and eighty days after emergence, and related these growth data to the amount of radiation intercepted by the crops. There is no evidence of any ontogenetic drift in the values of ε^* that can be calculated from their data until their last harvest. Using the factor 0.45 to convert total radiant energy to light energy, values of ε^* of about 3.5 $\mu g\ \mathrm{J^{-1}}$ can be calculated from their data. They also report the results of regressions of the standing total crop dry matter on cummulated intercepted radiation, and from these data average values of ε^* throughout growth of 3.5 and 3.7 can be calculated. These estimates of ε^* are far higher than those of ε_T given in Table 4.4. Their data can also be used to calculate the efficiency of light use in tuber production, which can be estimated as 3.3 μg (tuber dry matter) J^{-1}, suggesting that after tuber initiation about 90% of the new crop dry matter is partitioned to the tubers. Similar, high estimates of ε^* can be obtained for sugar beet (4.0 μg J^{-1}), barley (3.5 μg J^{-1}) and wheat (5.1 μg J^{-1}) from the data reported by Biscoe and Gallagher (1977). These high values may arise because the crops were fairly open arrays of plants. As we have seen in the previous section, if the product $\alpha k \bar{I}_0$ in eqn (4.21) were far less than F_{max}, the equation would reduce to

$$\varepsilon = q(Y_g \alpha - b). \tag{4.24}$$

If we neglect b, we can estimate an upper limit to the value of ε. Leaf photochemical efficiencies are usually about 12 μg (CO_2) J^{-1} (e.g. see Acock *et al.*, 1978), Y_g is about 0.75 (Hunt and Loomis, 1979) and q is about 0.71 (see Table 4.3). These values lead to an estimate of the upper limit of ε of around 6.4 μg (dry matter) J^{-1}.

Estimates of ε^* for a number of contrasting crop species can be derived from Table 5 in Warren Wilson (1971). These are shown here in Table 4.5. The important point to be made here is that these differences in the crop light-utilization efficiencies may be due as much to differences in crop management strategy as to the intrinsic genetic differences between plants.

Table 4.5

Estimates of the net light-utilization efficiency (ε^*) for different crops. (Derived from Warren Wilson, 1971).

Crop	ε^*/μg (dry matter) J^{-1}
Oryza sativa (rice)	4.2
Zea mays (corn)	3.4
Ipomoea batatas (sweet potato)	3.1
Brassica oleracea (Kale)	2.7
Helianthus annuus (sunflower)	2.6
Gossypium hirsutum (cotton)	2.5
Trifolium subterraneum (clover)	1.6
Glycine max (soybean)	1.3

Effects due to differences in crop management strategies (for example row widths, planting densities, etc.) are more appropriately investigated in the first instance in the field. It is easier to measure the amount of light intercepted by a row crop directly than to simulate it, and the measure is more certain. (see Section 3.6). However, it would also be valuable to make some measure of the light distribution within the canopy, so that effects due to differences in the distribution of light over leaf surfaces could be explicitly identified and perhaps accounted for.

5

Dry-Matter Partitioning

5.1　Root–shoot interactions

Synopsis: The growth of either a shoot or a root is dependent upon the activities of both the shoot and root. It has been suggested, on semi-empirical grounds, that root and shoot activities are proportional to one another. By suitable definition of these activities it can be formally shown that

> root mass × specific root activity
>
> = shoot mass × specific shoot activity

The derivation of this relationship is a tautology, but it provides a defined basis for studying the response of plant growth to different nutrient levels.

Luckwill (1960), Troughton (1960), Davidson (1969) and Hunt (1975) have all suggested, on semi-empirical grounds, that root and shoot activities are proportional to one another. Davidson (1969) has formalized this hypothesis by suggesting the relationship

root mass × specific root activity = shoot mass × specific shoot activity.

87

Application of this relationship to a variety of data sets can be found in the publications noted above.

The specific activity of an organ is defined as its activity expressed per unit mass of the organ, so that the specific root activity with respect to nitrogen uptake would be the rate of nitrogen uptake per unit of root mass. Similarly the specific activity of the shoot with respect to carbon uptake would by the rate of carbon assimilation per unit of shoot mass. The elemental composition of an increment in new plant dry matter, ΔW, with respect to some element M, f_M, is simply defined by

$$f_M = \Delta M / \Delta W \tag{5.1}$$

and if we denote the dry weight of the organ responsible for the assimilation of M by the symbol W_M, the specific activity of that organ with respect to the uptake of M, σ_M, is defined by

$$\sigma_M = (1/W_M)\,\Delta M / \Delta t. \tag{5.2}$$

We can use eqn (5.1) to substitute for ΔM in eqn (5.2), and rearrange the resulting relationship, when we obtain

$$f_M / \sigma_M = W_M / (\Delta W / \Delta t). \tag{5.3}$$

Roots take up a number of different minerals from the soil. The root mass, and its specific activity with respect to any one of the elements assimilated by it, can be denoted by the subscript R. Similarly, we can denote the shoot mass, and its specific activity, by the subscript T. If we now suppose that the element M is taken up by the roots, eqn (5.3) can be written as

$$f_M / \sigma_R = W_R / (\Delta W / \Delta t). \tag{5.4a}$$

Likewise, if we are considering the element carbon (C), which is taken up by the shoot, we can similarly write

$$f_C / \sigma_T = W_T / (\Delta W / \Delta t). \tag{5.4b}$$

Equations (5.4a,b) lead directly to the identity

$$W_T \sigma_T / f_C = W_R \sigma_R / f_M \tag{5.5}$$

and this is no more than a formal statement of Davidson's hypothesis. We have shown, without approximation, that Davidson's hypothesis is a logical consequence of the way in which we define root and shoot activities. The derivation of eqn (5.5) is a tautology, but nevertheless it is an important one. It has arisen because we have unambiguously defined root and shoot activities, and those definitions provide a basis on which we can explore the consequences of the functional relationship between root and shoot activities.

The total dry weight, W, of a vegetative plant is the simple sum of the root and shoot components; that is $W = W_R + W_T$. If we sum eqns (5.4a) and (5.4b), we obtain the relationships

$$f_M/\sigma_R + f_C/\sigma_T = (W_R + W_T)/(\Delta W/\Delta t) = W/(\Delta W/\Delta t). \tag{5.6}$$

We can now take reciprocals of both sides of eqn (5.6) so that it becomes

$$(1/W)[\Delta W/\Delta t] = \sigma_T\sigma_R/(f_C\sigma_R + f_M\sigma_T). \tag{5.7}$$

The specific growth rate of the plant, μ, is the rate of dry-matter change per unit plant weight and is formally defined by $(1/W)[\Delta W/\Delta t]$, so that eqn (5.7) leads directly to the relationship

$$\mu = \sigma_T\sigma_R/(f_C\sigma_R + f_M\sigma_T). \tag{5.8}$$

Using simple algebra and our definitions of root and shoot activities, we have described the plant specific growth rate in terms of the specific root and shoot activities and the elemental composition of the plant dry matter.

Let us now suppose that n different elements of interest to us are being assimilated by the roots. Equation (5.6) can then be written as

$$\sum^n (f_M/\sigma_R) + f_C/\sigma_T = (nW_R + W_T)/(\Delta W/\Delta t). \tag{5.9}$$

If we now replace $(\Delta W/\Delta t)$ in eqn (5.9) by μW, and rearrange the resulting expression, we have

$$\mu = (W_T + nW_R)\bigg/ W\bigg[\sum^n (f_M/\sigma_R) + (f_C/\sigma_T)\bigg]. \tag{5.10}$$

(Note that σ_R may be different for each of the n elements M.) We can explore the implications of eqn (5.10) by considering a particular application of it. For example, suppose that we are concerned with the nitrogen and phosphorus nutrition of a plant. We can denote the specific root activities with respect to nitrogen and phosphorus uptake by σ_{RN} and σ_{RP}, which enable us to re-write eqn (5.10) as

$$\mu = (W_T + 2W_R)/W[f_N/\sigma_{RN} + f_P/\sigma_{RP} + f_C/\sigma_T]. \tag{5.11}$$

We can rearrange eqn (5.11) by multiplying the top and bottom of the right-hand side of the equation by the product $\sigma_{RN}\sigma_{RP}\sigma_T$, when we obtain

$$\mu = (W_T + 2W_R)\sigma_{RN}\sigma_{RP}\sigma_T/W[f_N\sigma_{RP}\sigma_T + f_P\sigma_{RN}\sigma_T + f_C\sigma_{RN}\sigma_{RP}]. \tag{5.12}$$

If we make the procrustean assumption that the rates of uptake of nitrogen and phosphorus, or more particularly the specific activities of the roots to nitrogen and phosphorus uptake, are directly proportional to the concentrations of available nitrogen and phosphorus in the soil, N and P, and that the specific activity of the shoot with respect to carbon assimilation is directly proportional to the light flux density incident upon the plant, I, we can write eqn (5.12) in the form

$$\mu = a_0 NPI/(a_1 N + a_2 P + a_3 I) \tag{5.13}$$

where a_0 to a_3 are constants. Equation (5.13) suggests that the specific growth rate of the plant will respond hyperbolically to changes in N, P or I, and predicts an interaction between the levels of N, P and I. It immediately suggests an analysis for the plant or crop response to applied fertilizers, and this analysis will be considered in the next section.

If we reexamine eqn (5.5), we can see that the root to shoot ratio of a plant, W_R/W_T, can be obtained from it as

$$W_R/W_T = f_M\sigma_T/f_C\sigma_R \tag{5.14}$$

and by defining the root and shoot partition coefficients, η_R and η_T, by

$$\eta_R = W_R/W = W_R/(W_R + W_T) \tag{5.15a}$$

and

$$\eta_T = W_T/W = W_T/(W_R + W_T) \tag{5.15b}$$

we can use eqn (5.14) to obtain the identities

$$\eta_R = f_M\sigma_T/(f_C\sigma_R + f_M\sigma_T) \tag{5.16a}$$

and

$$\eta_T = f_C\sigma_R/(f_C\sigma_R + f_M\sigma_T). \tag{5.16b}$$

Equations (5.16a,b) define the root and shoot partition coefficients as functions of the root and shoot activities of the plant and the elemental composition of plant dry matter. Again, our definitions have enabled us to derive, without approximation, explicit descriptions of the root and shoot partition coefficients.

The derivation of equations (5.16a,b) has been discussed by Charles-Edwards (1976, 1981) and Thornley (1977). They tell us that the partition of new dry matter between roots and shoots may change with time because any, or all, of σ_R, σ_T, f_M and f_C change with time. (But note that f_M and f_C are the elemental compositions of the current increment in dry matter, not the overall elemental compositions of the plant dry matter.) Equations (5.16a,b) give an indication of the possible causes of ontogenetic variations in dry-matter partitioning. If the specific activities of the shoots or roots change with time as the plant ages, we might expect changes in the partitioning of new dry matter. Similarly, changes in the elemental composition of new plant dry matter may affect partitioning.

5.2 Crop response to fertilizers

Synopsis: A relationship derived in the previous section (eqn 5.13) appeared to be relevant to the study of crop response to fertilizer applications. The specific growth rate of the plant was related hyperbolically to the available nitrogen and phosphorus concentrations and the light level incident upon the plant. This relationship can be generalized to describe a similar dependence of crop yield on available nutrient concentrations. This general relationship bears a striking resemblance to the empirical predictor for the fertilizer requirements of crops that is widely used by agriculturalists in the United Kingdom.

I have shown in the previous section how a simple, logical analysis of root–shoot interactions leads to a mathematical equation that appears to be relevant to the analysis of experimental data for dry-matter partitioning between roots and shoots (eqns 5.16a,b). In developing the mathematics, an expression which may help us to understand crop responses to fertilizer applications (eqn 5.13) was also obtained, and it is worthwhile exploring this relationship in a little more detail.

Three features are common to almost all sets of sequential harvest data for field-grown crops. Firstly, during the early stages of crop growth the total standing dry matter of the crop appears to increase exponentially with time. Secondly, this phase of "exponential" growth gives way to a "linear" growth phase, when the total standing dry matter increases as a simple linear function with respect to time; that is the crop growth rate remains constant. Thirdly, towards the end of growth, the growth rate starts to decrease with time, and the total standing crop dry weight reaches a maximum value from which it may subsequently decline. In general we can approximate the changing rate of crop dry-matter production by

$$dW/dt = a_0 W(1 - a_1 W) \tag{5.17}$$

where a_0 and a_1 are constants, and W is the total standing dry weight of the crop. Equation (5.17) is the logistic growth curve. It can be derived from physiological considerations if three assumptions are made:

  (i) the plant or crop is completely defined by its dry weight, and the system is defined by the single state variable W, which varies with time;
  (ii) growth occurs at the expense of some single substrate;
  (iii) the rate of growth is linearly proportional to the product of substrate and plant weight.

The derivation of the logistic growth function is succinctly discussed by Thornley (1976, Ch. 1, Section 4).

If we integrate eqn (5.17) over the time interval $0-t$ days, where the initial standing total crop dry weight at time $t = 0$ is W_0, we obtain the logistic growth curve

$$\ln\left[W(1 - a_1 W_0)/W_0(1 - a_1 W)\right] = a_0 t \tag{5.18}$$

which can be rearranged into the form

$$W = W_0 \exp(a_0 t)/\{1 - a_1 W_0[1 - \exp(a_0 t)]\}. \tag{5.19}$$

The curve described by eqn (5.19) is illustrated in Fig. 5.1, where changes in the total standing dry matter of a crop are simulated using values of $W_0 = 0.01$ t ha^{-1}, $a_0 = 0.15$ d^{-1} and $a_1 = 0.1$ ha t^{-1}.

Now, we can write the specific growth rate of the crop, μ, as

$$\mu = (1/W)\, dW/dt = a_0(1 - a_1 W). \tag{5.20}$$

If we replace $(1 - a_1 W)$ in eqn (5.18) by μ/a_0 we can re-write it as

$$\ln\left[W(1 - a_1 W_0)a_0/\mu W_0\right] = a_0 t \tag{5.21}$$

which we can then rearrange to give

$$W = \mu W_0 \exp(a_0 t)/a_0(1 - a_1 W_0). \tag{5.22}$$

Equation (5.22) describes a relationship between the total standing dry weight of a crop and its specific growth rate. We can re-write eqn (5.22) more simply as

$$W = \mu f(t) \tag{5.23}$$

where

$$f(t) = W_0 \exp(a_0 t)/a_0(1 - a_1 W_0). \tag{5.24}$$

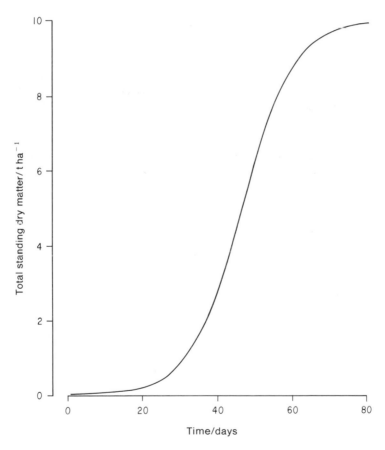

Fig. 5.1. The plant/crop growth curve described by eqn (5.19) (see text) with $W_0 = 0.01 \text{ t ha}^{-1}$, $a_0 = 0.15 \text{ d}^{-1}$ and $a_1 = 0.1 \text{ ha t}^{-1}$.

But we have seen that we can also write the specific growth rate, μ, as

$$\mu = (W_T + nW_R) \Big/ W\left[\sum_{n} (f_M/\sigma_R) + (f_C/\sigma_T)\right] \qquad (5.10)$$

so we could re-write eqn (5.23) as

$$W = \left\{(W_T + nW_R) \Big/ W\left[\sum_{n} (f_M/\sigma_R) + (f_C/\sigma_T)\right]\right\} f(t). \qquad (5.25)$$

If we consider only the effects of nitrogen fertilizer applications on total dry-matter production, eqn (5.25) can be simplified to

$$W = f(t)/[(f_N/\sigma_R) + (f_C/\sigma_T)]. \qquad (5.26)$$

Let us now suppose that the specific root activity with respect to nitrogen is a linear function of the level of application of nitrogen fertilizer, N, so that we can make the simplifying approximation

$$\sigma_R = a_0 N \qquad (5.27)$$

and let us further suppose that the specific shoot activity is directly proportional to the incident light flux density, I, so that we can also write

$$\sigma_T = a_1 I. \qquad (5.28)$$

We can now use eqns (5.27) and (5.28) to re-write eqn (5.26) as

$$W = f(t)a_0 a_1 NI/(f_N I + f_C N). \qquad (5.29)$$

The standing dry-matter yield at any time t, which we can denote by $Y(t)$, is then given by

$$Y(t) = A_0 IN/(f_N I + f_C N) \qquad (5.30)$$

where $A_0 = a_0 a_1 f(t)$. It may seem strange that we are writing a time variable $f(t)$ into a constant, A_0. Since we are looking at the effects of different fertilizer treatments on yield, we are comparing the dry weight of crops at the same point in time. That is, the duration of growth of the different crops will have been the same, so that t in eqn (5.24) will be the same for all crops in the comparison. Provided a_0, a_1, and W_0 were the same for all crops, A_0 should be the same. If we examined the response of the standing crop dry weight to, say, nitrogen and phosphorus fertilizers, eqn (5.30) would become of the form

$$Y(t) = A_0 INP/(f_N IP + f_C NP + f_P IN).$$ (5.31)

The foregoing mathematical analysis is not definitive, it provides only an approximate description of the real response of crop yield to fertilizer applications. For example, we have implicitly assumed that the yield of the crop is proportional to its total standing dry matter.

Thornley (1978) has developed a similar, although more sophisticated analysis, of crop yield response to fertilizer applications, and his analysis, from which this one derives, can be used to provide a theoretical basis for the empirical, practical predictor for the fertilizer response of crops which has been developed by Greenwood and co-workers (Greenwood et al., 1974) in the United Kingdom. Their predictor is of the mathematical form

$$1/Y(t) = [1/A_0 + 1/A_1 N + 1/A_2 P + 1/A_3 K][1/(1 - A_4 N)]$$ (5.32)

where A_0 to A_4 are constants and N, P and K denote the concentrations of available nitrogen, phosphorus and potassium in the soil. The additional term $[1/(1 - A_4 N)]$ allows for the observed adverse effects of high available nitrogen levels on crop yield. The implications and applications of the fertilizer response predictor (eqn 5.32) are discussed by Greenwood in greater detail elsewhere (Greenwood, 1981).

5.3 Dry-matter partitioning during vegetative growth

Synopsis: The analyses developed in Section (2.1) can be used to examine differences in dry-matter partitioning during vegetative growth attributable to differences in the nutrient status of the rooting medium of plants. The effects of changes in partitioning attributable to changes in the plant's aerial environment, and particularly changes in the air temperature, are not so amenable to analysis. However, these effects can be studied directly by experiment. Ontogenetic changes in partitioning patterns may also be attributable to ontogenetic changes in the specific activities of roots and shoots with respect to the uptake or assimilation of particular elements.

Let us first consider the partition of new dry matter between the roots and shoots of young plants. Equations (5.16a,b) enable us to write the root and shoot partition coefficients (η_R and η_T) as functions of the elemental composition of the new increment in total plant dry matter and the specific activities of the roots and the shoots. We can examine these relationships experimentally. For example, seedlings of *Lolium multiflorum* were grown in controlled environment rooms in pots containing washed sand. The seedlings were regularly watered with solutions containing different amounts of nitrogenous fertilizer. When they were harvested, the differences in the nominal amounts of nitrogen provided to the seedlings were reflected in their root and shoot partition coefficients and in the proportions of elemental nitrogen in their dry matter. These differences in η_R, η_T and f_N are recorded in Table 5.1.

Table 5.1

The proportion of dry matter in shoots (η_T) and roots (η_R), and the proportion of elemental nitrogen in plant dry matter (f_N) in seedlings of *Lolium multiflorum* grown with different nominal levels of available nitrogenous fertilizer.

Nominal level of N/ppm	η_T	η_R	f_N/g (N) g (dry matter)$^{-1}$
8	0.37	0.63	0.05
23	0.46	0.54	0.09
68	0.53	0.47	0.12
203	0.63	0.37	0.18

Now, if we replace f_M in eqn (5.16b) by f_N, and then take reciprocals of both sides, we have

$$1/\eta_T = 1 + (f_C \sigma_R / f_N \sigma_T). \tag{5.33}$$

We can now rearrange eqn (5.33) to give us

$$(1 - \eta_T) f_N / \eta_T = f_C \sigma_R / \sigma_T \tag{5.34}$$

where σ_R is the specific root activity with respect to nitrogen uptake. We can use the data given in Table 5.1 to calculate the ratio $f_C \sigma_R / \sigma_T$. Using the rate of light-saturated photosynthesis per unit of leaf dry weight as an index of treatment-induced variation in σ_T, and assuming f_C remains constant, we can estimate the relative response of the specific root activity, σ_R^*, to different nominal nitrogen fertilizer application levels. This is illustrated in Fig. 5.2.

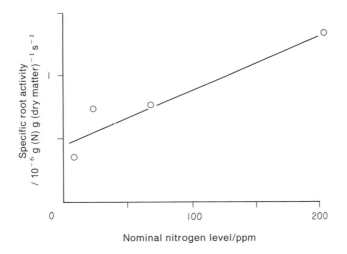

Fig. 5.2. The relative response of the specific root activity of seedlings of *Lolium multiflorum* to different nominal nitrogen fertilizer applications.

We have defined the root to shoot ratio (W_R/W_T) by eqn (5.14) as

$$W_R/W_T = f_M\sigma_T/f_C\sigma_R \tag{5.14}$$

and if we consider the effects of nitrogen on the root to shoot ratio (that is, we replace f_M by f_N), and assume that σ_R is proportional to the nominal nitrogen fertilizer level, N, (cf. Fig. 5.2), we can re-write eqn (5.14) as

$$W_R/W_T = f_N\sigma_T/f_C(a_0 + a_1N) \tag{5.35}$$

where a_0 and a_1 are constants. Provided f_N, f_C and σ_T remain constant, eqn (5.35) predicts a simple inverse relationship between the root to shoot ratio and the applied nitrogen level. The observed relationship between $1/(W_R/W_T) = W_T/W_R$ and the rate of supply of nitrogenous fertilizer for seedlings of *Diplacus aurantiacus* grown in controlled environment cabinets is shown in Fig. 5.3. These data, reported by Gulman and Chu (1981) were obtained for plants grown at the same nominal light level. There was little effect of different growth light levels in their experiment on the root to shoot ratio, suggesting that under their growing conditions σ_T was independent of the growth light level or there were compensatory changes in f_N, f_C or σ_R. Their data indicate that, as the growth light level increased, the rate of light-saturated photosynthesis per unit of leaf dry weight decreased (σ_T decreased) and that there was a concomitant decrease in the leaf nitrogen content (f_N).

The change in the rate of leaf photosynthesis was associated with marked changes in specific leaf area with the growth light level.

It seems unlikely that σ_R, σ_T and the elemental composition of new dry matter will remain unchanged during plant growth. If σ_R and σ_T change in the same direction (that is, if they increase or decrease together), the partition coefficients η_R and η_T may remain constant throughout growth. If they do not change together, we might anticipate ontogenetic changes in η_R and η_T.

The partition of dry matter between the vegetative parts of "closed" chrysanthemum crops growing in nutrient solution, but under natural daylight, has been studied by Acock and co-workers (Acock *et al.*, 1979). Whereas the proportions of dry matter in the different plant parts, roots and shoots, remained constant throughout the growth of the crops, their height

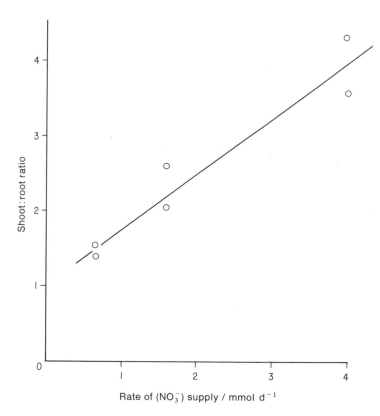

Fig. 5.3. The effect of different rates of nitrate (NO_3^-) supply on the shoot to root ratios of *Diplacus aurantiacus* (data from Gulman and Chu, 1981).

increased hyperbolically with time. The proportion of new dry matter partitioned to leaf and stem tissues did not change, implying that the leaf and stem densities (weight per unit volume of canopy) increased with time. These data, illustrated in Fig. 5.4, were collected over an eight-week period when the leaf area indices of the crop changed from about 2 to 10. It needs to be emphasized that there was no evidence of leaf, stem or root necrosis in these plants during the period of study.

Crops were grown at different temperatures at the same time of year (incident light level), and at the same temperature but different times of the year. Whereas the root and shoot partition coefficients, η_R and η_T, changed little with the different light levels, they were affected by the growth temperature. The root partition coefficient increased with increasing growth temperatures. The concomitant decrease in η_T was largely attributable to a marked decrease in the proportion of dry matter partitioned to stem tissues at the higher growth temperatures. These data are shown in Table 5.2. The fraction of elemental carbon in plant dry matter, f_C, appeared to increase with increasing growth light levels and decreasing growth temperatures.

For young, spaced plants growing under "steady-state" conditions we can show that

$$(1/W)\,\Delta W/\Delta t = (1/W_R)\,\Delta W_R/\Delta t = (1/W_T)\,\Delta W_T/\Delta t \qquad (5.36)$$

which is put more simply as

$$\mu = \mu_R = \mu_T \qquad (5.37)$$

where μ_R and μ_T are the root and shoot specific growth rates. This identity is illustrated in Fig. 5.5, using the data reported by Potter and Jones (1977) for the growth of young plants of nine species at different temperatures. The leaf specific growth rate, μ_L, is equal to the whole-plant specific growth rate across both species and treatments. Moreover we can write the leaf area of the plant, A, as

$$A = F_A W \qquad (5.38)$$

where F_A is the leaf-area ratio of the plant (leaf area to total plant dry weight). If we use eqn (5.38) to replace W in eqn (5.36) by A/F_A, we can easily show that the leaf-area specific growth rate, μ_A, should be identical to the plant specific growth rate. This is illustrated also in Fig. 5.5.

A comparison of the specific growth rates of different plant parts only tells us whether or not plants are growing under "steady-state" conditions, when losses of dry matter through any cause can be neglected. However,

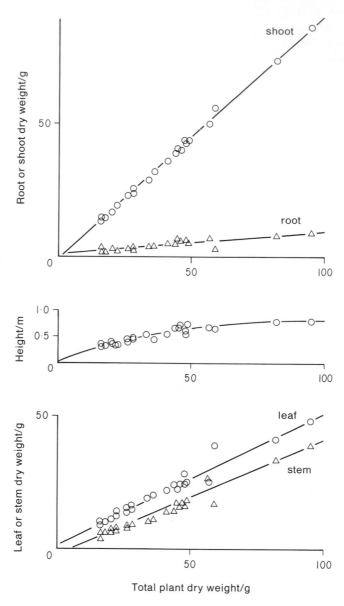

Fig. 5.4. Changes in the amounts of root, shoot, stem and leaf tissues on vegetative chrysanthemum plants with total plant weight. Plant height changes are also shown.

if we calculate the "modified" specific growth rate, that is the growth rate per unit of leaf dry weight (cf. eqns (2.7a,b,c) in Chapter 2) we can derive some information about partitioning. Let us denote the modified root and shoot specific growth rates here by μ_R^* and μ_T^*, then

$$\mu_R^* = (1/W_L)\,\Delta W_R/\Delta t \tag{5.39a}$$

and

$$\mu_T^* = (1/W_L)\,\Delta W_T/\Delta t \tag{5.39b}$$

where W_L is the leaf dry weight. But we have shown previously in Chapter 2 that

$$\mu_R^* = \eta_R \varepsilon b s_A S \tag{5.40a}$$

and

$$\mu_T^* = \eta_T \varepsilon b s_A S \tag{5.40b}$$

(eqns (5.40a,b) hold in the absence of any tissue death), so that the ratio of the modified specific growth rates, μ_R^*/μ_T^*, is equivalent to the ratio of the partition coefficients, η_R/η_T.

Table 5.2

Root (η_R), shoot (η_T), leaf (η_L), and stem (η_S) partition coefficients and the fraction of elemental carbon in plant dry matter (f_C) for "closed" vegetative crops of chrysanthemum plants grown at different controlled temperatures (T) and daily light integrals (S). (After Acock *et al.*, 1979).

$T/^\circ$C	$S/\mathrm{MJ\,m^{-2}\,d^{-1}}$	η_R	η_T	η_L	η_S	$f_C/\mathrm{g\,(C)\,g}$ $(\text{dry matter})^{-1}$
10	5.4	0.10	0.90	0.45	0.45	0.39
15	5.4	0.12	0.88	0.45	0.43	0.38
20	5.4	0.12	0.88	0.46	0.42	0.37
25	5.4	0.15	0.85	0.46	0.39	0.36
30	5.4	0.15	0.85	0.49	0.36	0.36
20	1.9	0.16	0.84	0.46	0.38	0.35
20	4.1	0.10	0.90	0.47	0.43	0.37
20	5.4	0.12	0.88	0.46	0.42	0.37
20	8.5	0.12	0.88	0.46	0.42	0.38
20	9.2	0.10	0.90	0.47	0.43	—

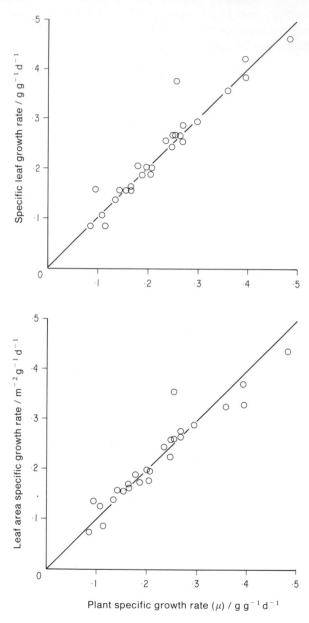

Fig. 5.5. The common identity of specific leaf growth rate (μ_L) total plant specific growth rate (μ) and leaf area specific growth rate (μ_A) (data from Potter and Jones, 1977).

5.4   Dry-matter partitioning during reproductive growth

Synopsis: When a crop becomes reproductive there are essentially two important considerations. Firstly, how many reproductive organs (pods or fruits) are set per unit of ground area by the crop? Secondly, what is their average size? It can be shown that provided some critical rate of supply of assimilate is required during the initiation of each reproductive unit, the numbers of reproductive units initiated per unit of ground area will be inversely related to the plant density. This relationship holds for soybean and mungbean crops.

Some plants, such as the tomato, behave in an agriculturally indeterminate way. After commencement of flowering and fruit set, they partition an almost constant proportion of their dry matter to the reproductive plant parts, the remainder being partitioned to new vegetative growth. Other plants behave in an agriculturally determined way. On flowering, they cease vegetative growth, that is new leaf production. I am only going to deal with those plants which behave in an agriculturally determined way.

Let us consider the growth of the above-ground parts of a crop after the cessation of vegetative growth. The rate of growth of these parts, $\Delta W_T/\Delta t$, can be written as

$$\Delta W_T/\Delta t = \eta_T \varepsilon J - V_T. \tag{5.41}$$

It is convenient if we change the parameters of eqn (5.41) by writing

$$\Delta W_T/\Delta t = a'_T Q - V_T \tag{5.42}$$

where a'_T is the maximum gross rate of above-ground dry-matter production when the crop is intercepting all of the incident light energy and Q is the actual proportion of the energy that is being intercepted by the crop. If we denote the harvestable, reproductive part of the crop by W_H (grain, seed or fruit), and ignore any losses of that component, we can write,

$$\Delta W_H = a_H Q - V_H \cong a_H Q \tag{5.43}$$

where $a_H = \eta_H a'_T$, and η_H is the proportion of new above-ground dry-matter that is partitioned to the harvestable part of the crop (since we have assumed that V_H can be ignored). Now let us suppose that for each unit of the harvestable component (that is each seed pod, ear of grain or fruit) assimilate needs to be provided at some critical rate during its initiation if it is not to be aborted. We can denote this critical rate of supply of assimilate by A_C, when the number of units of the harvestable component initiated per unit ground area, n, can be calculated as

$$n = a_H Q / A_C. \tag{5.44}$$

If the average number of harvestable units per plant is denoted by $\bar{n}$, we can also write

$$\bar{n} = n/\rho \tag{5.45}$$

where ρ is the planting density (that is the number of plants per unit of ground area). Combining eqns (5.44) and (5.45) then provides us with the simple expression

$$\bar{n} = a_H Q / \rho A_C \tag{5.46}$$

which predicts an inverse relationship between the average number of harvestable units per plant and the planting density. This relationship is observed between the average number of pods per plant in soybean crops (Enyi, 1973) and mungbean crops (Muchow and Charles-Edwards, 1982b) grown at different planting densities and the plant density. These data are illustrated in Fig. 5.6. The hypothesis that is represented by eqn (5.44) can be no more than a crude representation of reality. For instance, initiation of the harvestable units most probably occurs in all crops over a finite period of time. It is possible that the initiation of each harvestable unit depends upon the supply of assimilate only during some early critical period.

We can explore the relationship predicted by eqn (5.44) experimentally by manipulation of the crops during the initiation period. For example, we can shade crops (reducing a_H through a reduction in the daily incident light integral) or thin them (altering Q and ρ), and calculate the effect of these manipulations on $\bar{n}$. This has been done with a mungbean crop, and the comparison of predicted and observed pod numbers per plant is shown in Fig. 5.7 (Muchow and Charles-Edwards, 1982b). These data support the hypothesis formalized by eqn (5.44).

Whilst these analyses provide us with an insight into the effects of plant population density on the number of reproductive organs produced per plant, and thence the number per unit of ground area, they do not tell us what proportion of the current assimilate is partitioned to the reproductive organ. For the mungbean crops described here, between 57% and 78% of the current assimilate was considered to have been partitioned to the pods during reproductive growth. About 43% of the current assimilate was subsequently partitioned to the seed component of yield.

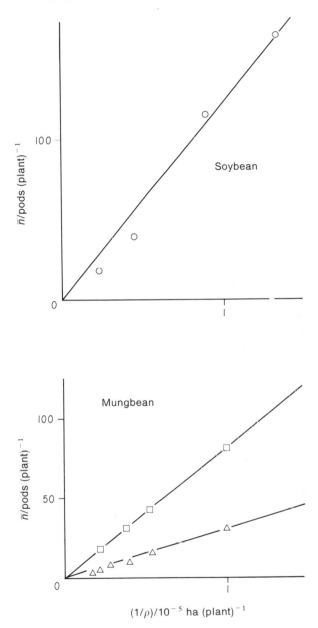

Fig. 5.6.   The relationships observed between the reciprocal of plant density $(1/\rho)$ and the mean number of pods per plant $(\bar{n})$ for mungbean crops (after Muchow and Charles-Edwards, 1982b) and soybean (after Enyi, 1973).

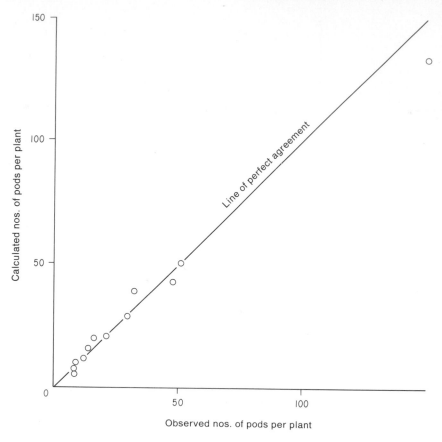

Fig. 5.7. A comparison of predicted and observed number of pods per plant following shading and thinning of mungbean crops during pod set (after Muchow and Charles-Edwards, 1982b).

5.5   Harvest index

Synopsis: The dependence of harvest index on phenological, physiological and environment effectors of crop yield can be explicitly defined. By making simplifying assumptions, specific relationships between some plant characters and harvest index can be obtained. There are considerable dangers of artificial correlations between harvest index and grain yield, and these can be simply illustrated. It is concluded that the harvest index adds little to our understanding of crop performance, that a better and more meaningful understanding can be obtained by the direct analysis of the main determinants of grain yield.

The harvest index of a crop is the ratio of its grain yield to the total, standing, above-ground dry matter of the crop at harvest. The total above-ground dry matter of the crop is often called its *biological* yield. If we denote the grain yield at harvest by W_G and the biological yield by W_T we can formally write

$$\text{harvest index} = W_G/W_T. \qquad (5.47)$$

There is a considerable body of experimental data showing that there is a strong association between harvest index and grain yield (e.g. Nass, 1973; Riggs *et al.*, 1981; Singh and Stoskopf, 1971), and the use of measurements of harvest index and biological yield have been advocated as an instrument that would make an important contribution to our understanding and advancement of crop performance (Donald and Hamblin, 1976).

Whereas harvest index is a simple parameter to obtain, the regression of grain yield (W_G) on harvest index (the ratio W_G/W_T) can be misleading and dangerous. If the variation in W_T is much less that that in W_G the regression can lead to an artificial correlation, and reports that "the increased yield of more recent varieties of oats has been associated almost entirely with increased harvest index, with no significant change in biological yield" (Donald and Hamblin, 1976) have little meaning. The potential fallacy of significant correlations between yield and harvest index is easily demonstrated. Austin and co-workers (Austin *et al.*, 1980) have examined the genetic improvements in winter wheat yields since the year 1900. In one of their experiments, with twelve wheat cultivars, they reported grain yields of between 3 and 5 $t\,ha^{-1}$ and biological yields of between 9 and 11 $t\,ha^{-1}$. Random grain and biological yields between these limits can be simulated for "twelve cultivars" using the random number generator on most small, programmable calculators. The linear regression of grain yield (W_G) on the derived harvest index (W_G/W_T) for one such set of "twelve random cultivars" is illustrated in Fig. 5.8. It demonstrates a high positive artificial correlation ($r^2 = 0.96$) between grain yield and harvest index. The problem of artificial correlations of this sort is elegantly discussed by Riggs 1970, Section 4.10).

The harvest index integrates phenological, physiological and environmental effectors of crop yield. Unless its dependence on each one of these categories of effectors can be explicitly defined, it is difficult to see how it can help us in developing any sort of understanding of crop performance. The net amount of above-ground dry matter produced by a crop during the interval $0-t$ days, its biological yield, can be obtained from eqn (1.3) as

$$W_T = \sum_{i=t} (\eta_T QS\varepsilon - V_T)_i \qquad (5.48)$$

where it is assumed that the initial standing dry weight of the crop is zero or close to zero.

Let us suppose that the crop grows vegetatively for a period of Δt_V days, and produces grain for a period of Δt_G days. The amount of grain produced during the period Δt_G days, W_G can be written as

$$W_G = \sum_{}^{j=\Delta t_G} (\eta_G S\varepsilon)_j \tag{5.49}$$

where η_G is the proportion of the daily increment of crop dry matter partitioned to the grain during grain production, and it is assumed that the daily losses of grain dry matter, V_G, can be neglected. The total standing crop dry matter at harvest will have two components, one produced during vegetative growth (Δt_V) and the other during grain production (Δt_G). We could therefore re-write eqn (5.48) as

$$W_T = \sum_{}^{i=\Delta t_V} (\eta_T Q S\varepsilon - V_T)_i + \sum_{}^{j=\Delta t_G} (\eta_T Q S\varepsilon - V_T)_j \tag{5.50}$$

which distinguishes between the parts produced during vegetative (Δt_V) and

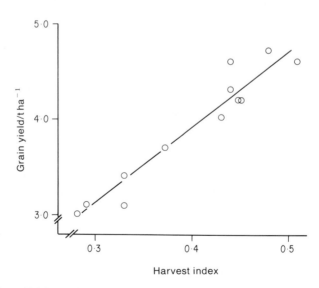

Fig. 5.8. The artificial correlation between grain yield and harvest index for twelve hypothetical cultivars of wheat. Grain yield and biological yield, for each cultivar, were produced by a random number generator (see text).

reproductive (Δt_G) growth. The harvest index of the crop can then be written as

$$\text{harvest index} = \sum^{j=\Delta t_G} (\eta_G QS\varepsilon)_j \Bigg/ \left[\sum^{i=\Delta t_V} (\eta_T QS\varepsilon - V_T)_i + \sum^{j=\Delta t_G} (\eta_T QS\varepsilon - V_T)_j \right].$$

$$(5.51)$$

Equation (5.51) describes the harvest index as an explicit function of phenological (Δt_V and Δt_G) and physiological (η_G, Q, ε, η_T, V_T) characters of the crop and the daily light integral incident upon it. Other environmental and physiological effectors are implicit through their effects on these eight determinants of harvest index.

Equation (5.51) is cumbersome, and its implications are difficult to envisage. In understanding these implications, it is helpful for us to make some procrustean assumptions about crop growth. So that we are aware of what we have assumed, it is useful to list the assumptions. Let us assume:

(i) that from the start of grain production no further dry matter is partitioned to stem and leaf tissue, and that all new dry matter during this period is partitioned to grain or chaff;

(ii) that the only above-ground tissues lost from the crop are the leaves, and that at harvest there are no live leaves remaining on the stem;

(iii) that there are no ontogenetic changes in ε throughout the entire growth of the crop, nor in η_S, η_G or η_C during the vegetative and reproductive growth periods (the subscripts S, G and C denote stem, grain and chaff respectively);

(iv) that the daily light integral, S, remains the same throughout growth of the crop.

These assumptions allow us to write

$$W_T = (\eta_S \bar{Q}_V \Delta t_V + \eta_G \bar{Q}_G \Delta t_G + \eta_C \bar{Q}_G \Delta t_G)\varepsilon S \qquad (5.52)$$

and

$$W_G = \eta_G \bar{Q}_G \Delta t_G \varepsilon S \qquad (5.53)$$

where $\bar{Q}_V$ and $\bar{Q}_G$ denote the proportions of the incident light energy intercepted by the crop throughout the vegetative and reproductive (grain filling) periods of growth, respectively. We can then use eqns (5.52) and (5.53) to re-write eqn (5.51) as

$$\text{harvest index} = \eta_G \bar{Q}_G \Delta t_G / [\eta_S \bar{Q}_V \Delta t_V + (\eta_c + \eta_G) \bar{Q}_G \Delta t_G]. \quad (5.54)$$

Equation (5.54) suggests that the harvest index will be hyperbolically dependent on the proportion of the light energy intercepted during grain production, the duration of grain production and the proportion of the dry matter partitioned to the grain during grain production. It also suggests that it will be inversely related to the proportion of the incident light intercepted during vegetative growth, the duration of the vegetative-growth period, the proportion of dry matter partitioned to stems during vegetative growth and the proportion of the dry matter partitioned to chaff during grain production.

If we denote the harvest index by H, we can rearrange eqn (5.54) to give

$$(1/H) - 1 = (\eta_S \bar{Q}_V \Delta t_V / \eta_G \bar{Q}_G \Delta t_G) + (\eta_c / \eta_G). \quad (5.55)$$

Now, Riggs and co-workers (Riggs *et al.*, 1981) have compared the yields of spring barley varieties grown in England and Wales between 1880 and 1890. If we assume that the ratio $\eta_S \bar{Q}_V / \eta_G \bar{Q}_G$ was reasonably constant across their different varieties, and that the duration of grain production, Δt_G, was also constant, we might anticipate a positive correlation between $(1/H) - 1$ and Δt_V for their cultivars. The observed association between $(1/H) - 1$ and Δt_V is illustrated in Fig. 5.9. Differences in the durations of the vegetative growth phases of the cultivars could account for some of the improvement in harvest index. The trend of a decreasing duration of vegetative growth (Δt_V), together with increasing harvest index (H), with the date of introduction of new cultivars is illustrated in Fig. 5.10.

Austin and co-workers have examined the genetic improvements in winter wheat yields in the United Kingdom since 1900, and the associated changes in the physiological characters of wheat genotypes introduced since 1900 (Austin *et al.*, 1980). They have concluded that the newer, higher-yielding varieties are shorter and reach anthesis earlier than the older varieties. They observed that the increase in grain yield of the newer cultivars was associated mainly with an increase in the harvest index of these cultivars. They argued that by continuation of the trend towards reduced stem length, with no change in above-ground biomass, breeders may be able to increase harvest index from the present value of about 50% to one of about 60%. As the limit to improvements in harvest index is approached, they argued, genetic gain in yield would depend upon exploiting genetic variation in biomass production. They claimed that their results showed that the yield improvements achieved by breeding new varieties resulted from morphological and physiological changes in the varieties in addition to those changes associated with increased lodging resistance. These data have been discussed previously in Section 2.4, and it was demonstrated that the differences they observed in the yields of the varieties could be attributed to differences in the amounts of light energy intercepted by them during grain production.

Before the harvest index can be used as a measure of the efficiency of grain production, additional information is needed on the plant or crop characters such as Δt_V, Δt_G, Q and η_G. Observations of the type reported by Gardener and Rathjen (1975), and illustrated graphically by Donald and Hamblin (1976), which show the harvest index declining in those cultivars that reach anthesis later, could be interpreted using analyses such as eqn (5.54).

The harvest index is derived directly from the grain yield of a crop. It is not particularly helpful or meaningful to look for associations between harvest index and a measurement from which that index was derived, grain yield. The harvest index can be written as a function of both phenological and physiological characters of the crop, but the relationships are too cumbersome to manipulate without making a number of procrustean assumptions. Since some of those assumptions may be open to serious challenge,

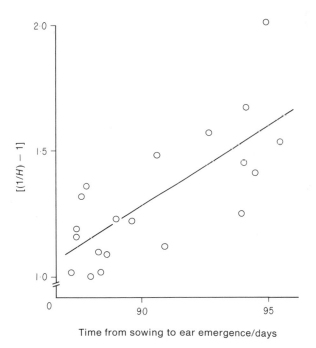

Fig. 5.9. The relationship between $(1/H) - 1$ and the period of time from sowing to ear emergence (Δt_V) for twenty barley cultivars. (H = harvest index).

the value of the relationships obtained is uncertain. It seems more logical, and the problems of improving grain yields more tractable, to look directly at the phenological, physiological and environmental determinants of grain yield.

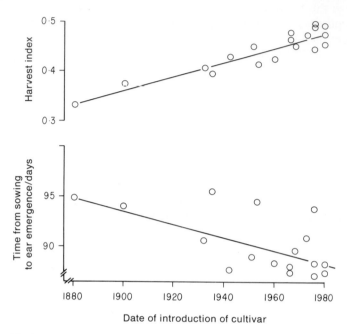

Fig. 5.10. The increase in harvest index (H) and decline in duration of the vegetative growth phase (Δt_V) of eighteen barley cultivars introduced into the UK since 1880.

6

Implications, Limitations and Applications

6.1 Introduction

Synopsis: The five physiological determinants of growth are amenable to direct estimation from field measurements, and three of them (the light-use efficiency, light interception and dry matter partitioning) can be related to our more basic, physiological understanding of crop growth. The analysis enables us to recognize three distinct phases of crop growth—establishment, vegetative and reproductive growth. There are discontinuous changes in one or more of the physiological determinants of crop growth between each of these growth phases.

In the first chapter a simple, logical analysis of the rate and extent of crop/plant dry matter production was developed. The analysis identified five determinants of crop dry matter production:

(i) the efficiency with which light energy is used in the production of new dry matter, denoted by ε and with units of g (dry matter) J^{-1} (light energy);

(ii) the amount of light energy intercepted by the crop/plant during each 24 h (day/night) period, denoted by J and with units of J (light energy) $m^{-2} d^{-1}$;

(iii) the proportion of the daily increment in *new* dry matter partitioned to the different crop/plant parts, denoted by η and dimensionless;

(iv) the daily rate of loss of dry matter, denoted by V and with units of $g\,m^{-2}\,d^{-1}$ (or $t\,ha^{-1}\,d^{-1}$);

(v) the duration of production of that part of the crop which is of interest, denoted by t, and with dimensions of time (days).

The *net* rate of *above-ground* dry-matter production, $\Delta W_T / \Delta t$, was written as

$$\Delta W_T / \Delta t = \eta_T \varepsilon J - V_T \qquad (6.1)$$

and the net amount of above-ground dry matter produced over t days, W_T, as

$$W_T = \sum^{i=t} (\Delta W_T / \Delta t)_i = \sum^{i=t} (\eta_T \varepsilon J - V)_i \qquad (6.2)$$

where the subscript T denotes parameters relating to the above-ground parts of the crop. It needs to be reemphasized that these five determinants of crop production are defined at the crop level of biological organization, not in terms of physiological processes occurring at some lower level of organization. If eqns (6.1) and (6.2) were used as the basis for a predictive, mathematical model of crop production, the model could be constructed with an empirical knowledge of the dependence of each of the determinants upon the cropping environment. However, it was shown in Chapters 3–5 how three of the determinants (ε, J and η) could be described as functions of lower-level, physiological processes. Equations (6.1) and (6.2) enable the analysis of crop growth data to be approached from two directions: the analysis of crop dry-matter production and the detailed physiological study of plant processes. If used as a basis for a crop simulation model, the analysis would meet Passioura's criterion that a model should be amenable to being tested at all levels of its hierarchical organization (Passioura, 1973). It would have the experimental advantage that the physiological understanding of each of the determinants of crop growth could be developed and tested separately.

The analysis described by eqns (6.1) and (6.2) does not replace traditional agronomic analyses of crop growth data, but rather complements them. If we can successfully describe crop production in terms of the five determinants defined here, we may more readily develop an *understanding* of the genetic, environmental and managerial effectors of growth through their effects on these determinants of growth. Such an understanding must inevitably affect our management of cropping systems. For example, if we have identified one of the determinants as being very susceptible to drought, we can use that knowledge to facilitate the design of crop-management strategies that will minimize the risk to the crop arising from that susceptibility.

There are important agricultural systems where this type of approach and analysis may be inappropriate. We need to be aware of these systems, which ones they are, and why the approach is neither satisfactory nor helpful to the analysis of them. Notwithstanding these limitations, the analysis may still provide a useful heuristic aid to our preliminary investigations of them. An example of the use of a simple mathematical analysis in this context is given in Section 6.3.

The application of this type of analysis to "field data" necessarily redirects us away from a descriptive experimental exercise towards one concerned with answering specific questions. It necessarily obliges us to be more precise in defining our experimental objectives. Casual inference might suggest that the application of this type of analysis would be time-consuming and demanding upon resources. That is not necessarily so. There are three distinct phases of crop growth. The first is the period between sowing and establishment of the crop, the second is the period of vegetative crop growth, and the third is the period of reproductive crop growth. The recognition of three distinct phases of growth defines four cardinal times in the life of the crop—sowing, establishment, commencement of reproductive growth and maturity. Although there may be ontogenetic or environmentally induced differences in some, or all, of the five determinants of growth during each of the three growth phases, there are discontinuous changes in one or more of them between phases. We can gain a great deal of understanding about these determinants by analysing the standing crop dry matter at these cardinal times. Extending the depth of our understanding of any one of these determinants may involve more intensive measurements during a particular growth phase. However, these more intensive measurements will be directed towards achieving well-defined objectives. We will have broken the resource-demanding cycle of collecting information because "it might be of use", and will be collecting only information that we can see is of use.

6.2 Extensive systems

Synopsis: The physiological analysis of crop growth data from systems which display marked, but poorly defined, heterogeneity in either their physical environment or their botanical composition (extensive cropping systems) is not readily undertaken. The analysis leads directly to the idea of a dynamic growth index to predict crop performance by these systems.

All the analyses that have been discussed in the preceeding chapters have either been based on the assumption that the systems being studied were spatially homogeneous with respect to both their botanical composition and their physical environment, or on the alternative assumption that any spatial heterogeneity could be defined and explicitly accounted for in the analysis (cf. Sections 3.4 and 3.5). However, many cropping systems display marked, but ill-defined, spatial heterogeneity in both their composition and their physical environment. Such systems include monocultures covering large land areas, intensively cultivated inter-cropping systems and pastoral grazing systems. The phrase "extensive cropping system" is used here to describe any system where there is marked spatial heterogeneity in composition or physical environment which can neither be neglected, nor can be simply and explicitly accounted for in the analysis of the system.

The marked spatial heterogeneity in the physical environment of extensive systems, the variations in their topography, soil type and fertility, etc. can complicate and confound the traditional statistical analyses of measurements which have been made on them. Let us suppose that we are researching an extensive system which displays marked spatial heterogeneity in soil types. Let us suppose that we can distinguish n distinct soil types, each with its own characteristic hydraulic properties and fertility. We are interested in predicting the yield response of the system to some independent variable X. If the yield within each of the n domains of soil, which we can denote by Y_i, responds linearly to changes in X, so that we can write

$$Y_i = a_i X + b_i \tag{6.3}$$

where a_i and b_i are constants which are characteristic of the soil type in the ith domain, the average yield over all domains, denoted by $\bar{Y}$, can be written as

$$\bar{Y} = \left(\sum_{i=1}^{i=n} Y_i \right) \Big/ n = \left(\sum^{n} a_i \right) X/n + \left(\sum^{n} b_i \right) \Big/ n. \tag{6.4}$$

If we denote the average soil characteristics over the whole region by $\bar{a}$ and $\bar{b}$, where

$$\bar{a} = \sum^{n} a_i/n \tag{6.5a}$$

and

$$\bar{b} = \sum^{n} b_i/n \tag{6.5b}$$

we can re-write eqn (6.4) as

$$\bar{Y} = \bar{a} X + \bar{b}. \tag{6.5c}$$

The average yield of the region, $\bar{Y}$, is described as a function of the independant variable X and the average soil characteristics of the region. But let us now suppose that the response of the yield to X is a non-linear function, so that

$$Y_i = a_i X/(1 + b_i X). \tag{6.6}$$

The average yield of the region, $\bar{Y}$, then becomes equal to

$$\bar{Y} = \sum_{i=n} [a_i X/(1 + b_i X)]/n \tag{6.7}$$

and we are not able to define simple average soil characteristics for the region. We can separately analyse the factors affecting yield in each of the n domains, and then compute the average yield, or we can seek an alternative approach. It is useful and relevant here to read Thornley's comments on how we can develop procedures for selecting a sample plant out of a population, where the sample plant is typical of the population in a defined way (see Thornley, 1976, Ch. 14). He discusses the problems of describing the heterogeneity of a population of plants, and establishing a procedure for selecting a "typical" single plant. The procedure he describes is meaningful only if the plant population is qualitatively homogeneous. For example, if half the population of plants were vegetative, and the other half were flowering, it would be necessary to first divide the population into two sub-groups, the vegetative and the flowering plants. If the soil characters of the region were not qualitatively homogeneous, we would similarly have to subdivide the region into sub-regions, where each sub-region was qualitatively homogeneous.

If we wish to predict, rather than analyse, the yield of an extensive cropping system, the complexities introduced by the explicit description of the spatial heterogeneity within the system may be overwhelming. An alternative approach is not to base the prediction on an analysis of the separate physiological determinants of the crop growth in the system, but to use primary yield data, averaged over the whole system, to integrate the heterogeneity for us. This approach leads directly to the idea of a growth index.

Essentially, the growth index is the ratio of the actual performance of the crop, averaged over the whole area, to its potential performance in the absence of any limitations imposed upon it by either the water or mineral nutrient deficiencies. The development of an index, however empirical, requires some model of the crop response to environment. In an unvarying and entirely reproducible environment, the growth index may be a static, scaling factor. However, in real cropping situations there is often a high association between the visible spatial heterogeneity in crop performance and stochastic temporal events such as rainfall or insect infestation. This association can be illustrated with a simple example. Water deficits in the soil, and thence water deficits in the plant, are likely to develop most rapidly in, and on, free-draining soils. Following an isolated fall of rain in an otherwise dry environment, we might expect to see spatial heterogeneity in crop performance associated with the spatial heterogeneity in the hydraulic properties of the soils in the area. However, if the periods of rainfall were frequent, and the amount of rain was adequate, soil water deficits would not develop on the free-draining soils, and plant performance would be homogeneous across the whole area despite the spatial heterogeneity in soil hydraulic properties. It may, therefore, be preferable, indeed necessary, to take into account the stochastic, temporal changes in the cropping environment by establishing a dynamic growth index.

A dynamic growth index introduces other sorts of problems if we wish it to have a well-defined basis in our physiological understanding of plant growth. The growth rate of a plant is a state-dependent variable. Do we define a growth index as the ratio of actual to potential plant growth rates at the same plant state (say at the same leaf area index), or do we attempt to define a state-independent potential plant growth rate (say the plant/crop growth rate at full light interception)? Whereas an effector of plant growth may affect any one determinant of growth, say the plant light-utilization efficiency, in a linear way, if it affects two or more of the determinants of plant growth it may affect their product, the plant growth rate, in a highly non-linear way. There is a powerful argument for abandoning any pretence of describing the growth index in detailed physiological terms. However, the growth index may be based, albeit empirically, on a model of the effects of environment on the physiological determinants of crop growth. Two examples of the "growth index" approach are illustrated below.

The first example is the use of an index of crop water stress, related to wheat and grain sorghum yields, to predict yields of these crops (Nix and Fitzpatrick, 1969). Intra-seasonal and year-to-year differences in varietal yields from both crops, when grown in central Queensland, appear to be largely a function of anthesis date in relation to the environmental water stress prevailing at that time. An index incorporating available water supply by the soil and potential evaporative demand was found to give highly significant correlations with grain yield, accounting for between 60% and 83% of the yield variation within individual wheat and grain sorghum varieties. The index was extended to several centres, accounting for 60–70% of total yield variation at each one, and 66% of the yield variation over the whole range of experimental centres. The best index was found to be the ratio between estimated values of available water in the root zone and the mean potential evapotranspiration at anthesis.

The second example, also based on water availability, deals with a pastoral system. McCown (1981a) has developed a simple dynamic growth index for forage dry-matter production for pastures in tropical Australia to investigate the potential for beef-cattle production in this area. The index, which is estimated on a weekly basis, is based on the availability of water for pasture growth and the mean daily temperature. Implicit in the index is the assumption that the temporal variation in the amount of water available for plant growth in arid and semi-arid areas of northern Australia is as great as, or greater than, the spatial variation due to heterogeneity in the soil hydraulic properties. Although empirical, the index is based on a simple model that the differences between rainfall and the evaporative losses of water within any week reflect the amount of water available for plant growth. Trends in the weekly growth index are found to be correlated with trends in animal live-weight changes. The growth index, which integrates environmental information directly into a predictor of animal live-weight changes, provides a basis for the examination of geographic and year-to-year variations in cattle live-weight gain (McCown, 1981b).

Let us return to more general aspects of the study of extensive systems. If we are to be able to analyse, and then predict, the behaviour of an extensive cropping system, we need to know about the dynamics of plant growth and the dynamic interactions between plants of different types. Because of the size and perceived complexity of these extensive systems, experimental research on them has often been limited to static descriptions of them. A treatment has been applied to the system, say stocking rate on a pasture has been increased, and the effects of the treatment have been assessed at a single point in time. Whilst these measurements must represent a real solution to the dynamic behaviour of the system, they provide us with little understanding of that behaviour. This can be illustrated quite simply. Consider the data collected on an extensive three-component system (perhaps a pasture) at the start, middle and end of the growing season (as shown in Table 6.1). Whereas plant species B remains a constant proportion of the total biomass of the system throughout the growing season, species A appears to be the dominant plant species in the middle of the growing season and species C the smallest component. Even if we record the absolute dry matter yields of one, or all, of the three species at each harvest, we gain little extra information. We cannot establish from these data whether or not the three species are behaving in a physiologically similar manner. The "data" shown in Table 6.1 were obtained from the simulated plant growth data illustrated in Fig. 6.1. For the purposes of this simulation the three species were assumed to display identical linear growth with respect to time, but their growth periods were displaced in time during the growing season. This simple exercise illustrates two points. Firstly, static measurements or observations of highly dynamic systems may provide us with little information and understanding about those systems. Secondly, systems comprised of

several, non-interactive components can often exhibit quite complex properties when viewed from outside. Table 6.1 and Fig. 6.1 can be taken to
illustrate the simple point that, if we can identify the separate components
of a system and study their individual behaviour *in situ*, we will gain a better
understanding of the behaviour of the complete system. This is really no
more than a restatement of the rationale outlined in Chapter 1, but applied
at a higher level of biological organization.

Table 6.1

Seasonal changes in the composition of a three-
component cropping system at the start, middle and
end of the "active" growth period. For species B, the
numbers in brackets are the standing dry weights in
arbitrary units.

	%Composition of standing dry-weight		
	Start	Mid-season	End
Species A	33	42	33
Species B	33 (0.5)	33 (2.0)	33 (0.5)
Species C	33	25	33

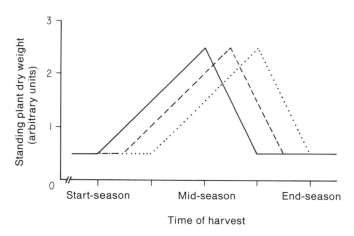

Fig. 6.1. Simulated dry matter growth curves for the three-component plant species of an
extensive cropping system (see text).

6.3 Plant competition: a heuristic, dynamic model of a binary cropping system

Synopsis: In general, models of the competition between plants in a mixed stand of plant types represent static descriptions of the mixture. They relate the yield of each plant type, at a particular moment in time, to the proportions of that plant type in the mixture. The inter-seasonal or intra-seasonal dynamics of the competition are dealt with by exploring the dynamic changes in the static descriptors.

A dynamic model of the growth of two different plant types when in competition with one another can provide a useful heuristic aid in helping us to understand the possible causes the consequences of competition between the two plant types.

Antonovics (1978) has examined three models of competition among plant species and has extended them to examine competition among interbreeding genotypes. He observes that theoretical studies invariably predict that the competitive relationships of two species will not be permanent, but that the species will diverge in resource requirements. The many experimental studies indicate that selection of genotypes for competitive relationships may produce variable results, since competitive ability is a complex trait. Competition models, such as de Wit's model (de Wit, 1960) are essentially static descriptions of plant competition. These models describe the yield (or standing biomass) of one plant component of a mixture of plants as a function of its yield when grown alone, and suggest that the competitive ability of the plant, its aggressivity, can be deduced from the departure of its yield in the mixture from the yield that might be expected if there were no competition between the plant types (McGilchrist and Trenbath, 1971). For example, these authors have proposed that the aggressivity, denoted by A_{ij}, can be written as

$$A_{ij} = [(Y_{ij}/Y_{ii}) - (Y_{ji}/Y_{jj})]/2 \qquad (6.8)$$

where Y_{ii} and Y_{jj} are the yields of the ith and jth genotype within the mixture when they are grown in monoculture, and Y_{ij} and Y_{ji} their yields when grown together (Trenbath, 1978). The relative crowding coefficient, k_{ij}, for species i grown in association with species j in a two-component mixture can be written as

$$k_{ij} = Y_{ij}Z_i/(Y_{ii} - Y_{ij})Z_j \qquad (6.9)$$

where Z_i and Z_j are the relative frequencies of the species i and j in the mixture (Hall, 1974). Equations (6.8) and (6.9) are introduced only to emphasize the essentially static nature of the descriptions. Dynamic changes in A_{ij} and k_{ij} could be studied experimentally, but they are empirical descriptors of the competition and provide no mechanistic understanding of its causes. The inter-seasonal or intra-seasonal dynamics of competition between plants are usually studied by looking at the dynamics of these types of static descriptor.

It can be a rewarding and stimulating exercise to construct a simple mathematical model of a cropping system before embarking upon any large-scale experimental investigations of it. The model can help the experimenter to define objectives for this program, and can often help him in deciding which measurements he should make and which observations he should record. The model becomes an heuristic aid to the researcher, that is it teaches him about the sort of behaviour that he might expect from the system. It is useful and instructive to illustrate this type of modelling exercise here.

Let us suppose that we are interested in the productivity and long-term stability of a self-seeding, extensive cropping system, perhaps a pasture. Let us further suppose that the system is composed of two annual plant species, A and B, that compete with each other for some resource during their growing seasons. Since the purpose of the model is to educate ourselves about the "likely" behaviour of the system, we want to use only the simplest mathematics to describe it. If we can obtain analytical solutions to the differential equations describing the temporal changes in the state of the system, we will be able to understand the subtleties of its behaviour better.

Firstly, let us suppose that the ontogenies of the two species are identical. We will assume that they germinate on the same day and reach maturity on the same day. Further, we will assume that the potential growth rates of their above-ground parts, a_T, are the same and that their actual growth rates, when they are growing together can be simply described by

$$dA/dt = a_T A/(A + B) \tag{6.10a}$$

and

$$dB/dt = a_T A/(A + B). \tag{6.10b}$$

We can rearrange eqns (6.10a,b) to obtain the identity

$$(1/A)\,dA/dt = (1/B)\,dB/dt \tag{6.11}$$

which on integration yields

$$A/A_0 = B/B_0 \qquad (6.12)$$

where A_0 and B_0 are the initial weights of A and B at time $t = 0$. Equation (6.12) can then be used to substitute for B in eqn (6.10a) or A in eqn (6.10b). This enables us to integrate eqns (6.10a,b) over the time interval $0-t$ to obtain the growth functions

$$A = A_0[1 + a_T t/(A_0 + B_0)] \qquad (6.13a)$$

and

$$B = B_0[1 + a_T t/(A_0 + B_0)]. \qquad (6.13b)$$

Since A and B germinate and reach maturity at the same time we can write

$$A_m/B_m = A_0/B_0 = A_s/B_s \qquad (6.14)$$

where A_m and B_m are the standing dry weights of A and B at maturity and A_0 and B_0 can be replaced by the viable seed weights per unit of ground area of A and B at germination, A_s and B_s. Finally, if we assume that the amount of viable seed at the start of the nth growing season is directly proportional to the mature standing dry weight of the crop at the end of the $(n - 1)$ growing season (that is $(A_s)_n = \beta_A(A_m)_{n-1}$ and $(B_s)_n = \beta_B(B_m)_{n-1}$, where β_A and β_B are the proportionality constants) the ratio of the amounts of viable seed of A and B at the start of the nth growth season, $(A_s/B_s)_n$, can be written as

$$(A_s/B_s)_n = (\beta_A/\beta_B)^{n-1}(A_s/B_s)_1 \qquad (6.15)$$

where $(A_s/B_s)_1$ is the ratio at the start of the first season.

Equation (6.15) is an important result. Since we have assumed that the ontogenies of A and B and their potential growth rates are identical, eqn (6.15) implies that if (β_A/β_B) is less than unity the plant species B will eventually dominate the system. This dominance of species B, which increases over successive seasons, is illustrated in Fig. 6.2. This figure shows dry-matter production over four successive seasons, simulated using eqns (6.13a,b), with values of the ratio (β_A/β_B) of 1.0, 0.5 and 0.25, with an initial value of $(A_s/B_s)_1$ of 2.0. If A and B are cultivars of the same self-seeding annual plant species, the proportion of dry matter partitioned to seed (β_A/β_B) could be a major determinant of the evolutionary dominance of one or other of the cultivars.

It is unlikely that the ontogenies of two different plant species will be identical, and we can explore the consequences of differences in their ontogenies on the stability and productivity of the system. Let us suppose that, although their growth rates are identical, species A germinates at some time, Δt days, before species B, and that the lengths of their growing seasons, t_A and t_B days, are also different. We can distinguish four distinct growth phases in this simple binary system.

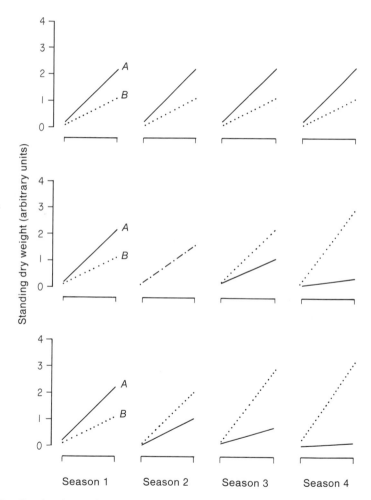

Fig. 6.2. Simulated growth curves for four successive seasons of the two-component plant species in a binary extensive cropping system. The species are assumed to have identical potential growth rates and ontogenies. The relative proportions of their dry weights at maturity partitioned to seed for the next season (β_A/β_B) are (a) 1.0, (b) 0.5 and (c) 0.25. At the start of the first growing season it was assumed that $(A_s/B_s)_1$ was equal to 2.0.

1. Growth of A prior to germination of B.   During this growth phase, which lasts for Δt days, there is no competition between species A and B, since B has not germinated. We can write the standing dry weight of A at the end of this growth phase, A_0, as

$$A_0 = A_s + a_T \Delta t \qquad (6.16)$$

(cf. eqn 6.13 with $B_0 = 0$), where A_s is the initial viable seed weight of A. The amount of B present will remain as B_s, the viable seed weight of B.

2. Growth of A and B until maturity of A.   The duration of this phase of growth will be $(t_A - \Delta t)$ days. Equations (6.13a,b) can be used to calculate the weights of A and B at the end of the growth phase as

$$A_m = A_0\{1 + [a_T(t_A - \Delta t)/(A_0 + B_s)]\} \qquad (6.17a)$$

and

$$B_0 = B_s\{1 + [a_T(t_A - \Delta t)/(A_0 + B_s)]\}. \qquad (6.17b)$$

We can combine eqns (6.16a,b) and (6.17a) to calculate the standing dry weight of A at maturity, A_m, as

$$A_m = [A_s + a_T \Delta t][A_s + B_s + a_T t_A]/(A_s + B_s + a_T \Delta t). \qquad (6.18)$$

From eqns (6.17a,b) we also have the relationship that

$$B_0 = B_s A_m/A_0 \qquad (6.19)$$

and if we use eqns (6.16) and (6.18) to substitute for A_m and A_0 in eqn (6.19) we can obtain

$$B_0 = B_s(A_s + B_s + a_T t_A)/(A_s + B_s + a_T \Delta t). \qquad (6.20)$$

3. Post-maturity loss of A and growth of B to maturity.   We will assume that after maturity the standing dry weights of both species decline linearly with time to zero, according to

$$A = A_m - \gamma t \qquad (6.21a)$$

and

$$B = B_m - \gamma t \qquad (6.21b)$$

where γ is a dry-matter loss constant. Now we can use eqn (6.21a) to substitute for A in eqn (6.10b) and write that the rate of growth of B during this phase is given by

$$dB/dt = a_T B/(B + A_m + \gamma t) \qquad 0 < A_m/\gamma \qquad (6.22)$$

whilst A is non-zero, and

$$dB/dt = a_T \qquad A_m/\gamma < t < t_B - t_A + \Delta t \qquad (6.23)$$

where A has become zero. The duration of this growth phase will be $(t_B - t_A + \Delta t)$ days, and the time taken for the standing dry weight of A to decline to zero will be (A_m/γ) days.

Equation (6.22) is integrable. If we make the substitution $Z = A_m - \gamma t$, then we also have that $dZ = -\gamma\, dt$. We can now re-write eqn (6.22) as

$$-\gamma(dB/dZ) = a_T B/B + Z \qquad (6.24)$$

and if we invert the equation, and rearrange it we can obtain

$$dZ/dB + \gamma Z/a_T B = -\gamma/a_T. \qquad (6.25)$$

Equation (6.25) is now a general linear, first-order differential equation which can be solved by the standard methods of elementary calculus. The solution to it is

$$A_m - \gamma t = -\gamma B/(a_T + \gamma) + \{[A_m(a_T + \gamma) + \gamma B_0]B_0^{(\gamma/a_T)}B^{(-\gamma/a_T)}\}/(a_T + \gamma) \qquad (6.26)$$

where I have replaced Z by the original term $A_m - \gamma t$. Now, if $t = A_m/\gamma$, eqn (6.26) simplifies to

$$0 = -\gamma B + [A_m(a_T + \gamma) + \gamma B_0]B_0^{(\gamma/a_T)}B^{(-\gamma/a_T)} \qquad (6.27)$$

which rearranges to

$$B = [B_0 + A_m(a_T + \gamma)/\gamma]^{\{a_T/(a_T + \gamma)\}}B_0^{\{\gamma/(a_T + \gamma)\}} \qquad (6.28)$$

and it is convenient if we denote this value of B at the time $t = A_m/\gamma$ by B^*. If A declines to zero before B reaches maturity, that is $(A_m/\gamma) < (t_B - t_A + \Delta t)$, we can then calculate B_m as

$$B_m = B^* + a_T(t_B - t_A + \Delta t - A_m/\gamma) \qquad (6.29)$$

where B^* is given by eqn (6.28). If B reaches maturity before the standing, above-ground dry weight of A has declined to zero, B_m can be calculated directly from eqn (6.24), with t equal to $(t_B - t_Z + \Delta t)$.

4. Post-maturity losses of B^*. When B attains maturity, its standing dry weight will decline with time according to eqn (6.21b). The duration of this growth phase will be B_m/γ days.

It needs to be reemphasized that this is an heuristic simulation model. It is not intended that it should describe any particular system, but is constructed only to provide us with some insight into the complex behaviour of real systems. For example, we can usefully use it to examine the effects of different lengths of time between germination and maturity for each species on the dry-matter composition of the whole system. We can assume that in a particular system species B germinates 5 days after species A ($\Delta t = 5$), and that whereas species A matures 40 days after germination, species B reaches maturity after either 45, 55 or 65 days. The potential growth rates, a_T, and dry-matter loss rates, γ, can be assumed to be the same for both species ($5 \text{ g m}^{-2} \text{ d}^{-1}$ and $20 \text{ g m}^{-2} \text{ d}^{-1}$ respectively). We can also assume that both species partition 10% of their mature dry weights to viable seed for the next season. Their growth, simulated during the tenth successive season, in each of those three cases (for $t_B = 45$, 55 and 65 days), are illustrated in Fig. 6.3. Whereas B has effectively been competed out by A when it only has 45 days to grow between germination and maturity, it survives quite successfully when t_B is assumed to be 55 days or longer. Changes in the total standing dry weight of the system are illustrated in Fig. 6.4. They illustrate the intuitive conclusion that increasing the duration t_B will increase the length of time that there is a substantial standing dry weight of plant material in the system.

We can compare these simulations with those obtained when we postulate different germination dates for the two species but unchanging lengths of time between germination and maturity. In these simulations t_A and t_B remain at 40 and 45 days throughout, but Δt is varied between 5 and 25 days. Again, the simulations have been run until they begin to converge on stable solutions, during the tenth season. Increasing Δt increases the amount of B produced at maturity and the total dry-matter productivity of the system (Figs. 6.5 and 6.6).

We can also interpret these simulations in another context. Suppose that A and B are two different genotypes of the same species. If their ontogenies are the same, they will both remain stable components of the population. If genotype B germinates later than genotype A, it will be competed out of the population unless it either increases the length of time it needs to mature or it delays germination long enough for it to be able to attain maturity after A has died. The simulations could illustrate selection pressures covering divergence of the ontogenies of the two genotypes.

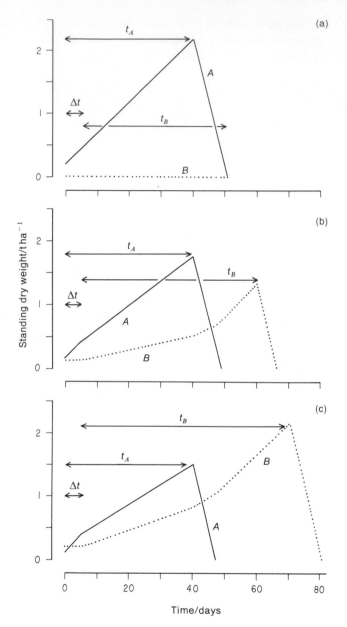

Fig. 6.3. Simulated growth during the 10th season of the two components of a binary extensive cropping system. In all simulations a_T and γ were the same for both plant species, and the difference in length of time between germination of the two species was 5 days in all simulations. The differences between growth of the two species in the simulations is due to the different lengths of time assumed to occur between the germination and maturity of species B. In simulation (a), $t_A = 40$ days and $t_B = 45$ days; in (b), $t_A = 40$ days and $t_B = 55$ days; in (c), $t_A = 40$ days and $t_B = 65$ days.

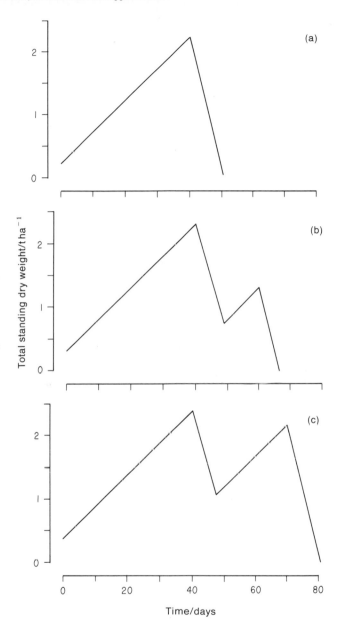

Fig. 6.4. Changes in the total dry-matter production of a binary extensive cropping system, during its 10th growing season, attributable to differences in t_B (see Fig. 6.3).

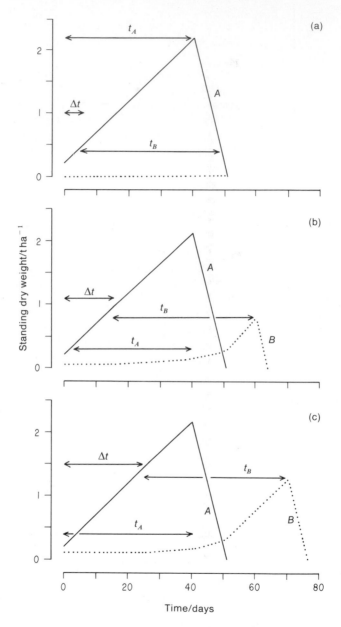

Fig. 6.5. Simulated growth during the 10th season of the two components of a binary extensive cropping system. The simulated differences between the two species are attributable to the increasing delay between germination of the seed of the two species (Δt). In simulation (a); $\Delta t = 5$ days; in (b), $\Delta t = 15$ days; in (c), $\Delta t = 25$ days.

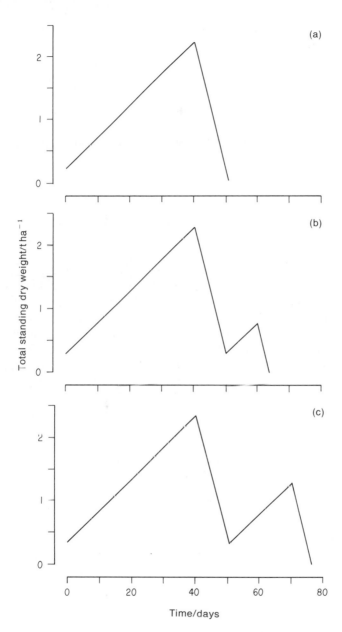

Fig. 6.6. Changes in the total dry-matter production of a binary extensive cropping system, during its 10th growing season, attributable to differences in Δt (see Fig. 6.5).

6.4   Applications to the analysis of crop data

Synopsis: The analysis described in the previous chapters can be applied directly to field crop data. However, some assumptions need to be made to do this, and they will differ for different crops. Firstly, the gross rate or amount of dry-matter production needs to be inferred from the net rate of dry-matter production or the net standing amount of crop dry matter. Secondly, the amount of light energy intercepted by the crop during the period of interest needs to be measured or estimated. A variety of measurements which will relate to physiological and environmental effectors of the physiological determinants of growth can be made on the crop.

Three phases of crop growth were described in Section 6.1—the establishment, vegetative and reproductive phases. I have deliberately avoided an unambiguous definition of the establishment phase of crop growth. It is that period between sowing and the emergence of the maximum numbers of viable seedlings. The identification of three phases leads to the consequent identification of four cardinal times during the life of the crop. The analyses developed in this book are confined to the vegetative and reproductive crop growth phases, and more particularly to the production of above-ground dry matter during these phases. The harvestable component of many crops is beneath the ground. It may be a below-ground storage organ, as in the potato, or a below-ground asexual reproductive organ, as in the onion crop or a bulb crop. The arguments used here can be readily extended to these crops, although it may be convenient to distinguish two sub-phases of vegetative growth for them, growth before and growth after the initiation of the below-ground organ.

Whereas the time of crop establishment is an almost entirely subjective quantity, the time of transition from vegetative to reproductive growth is more objective. It needs to be remembered that the crop is a collection of individual plants. The transition of these individuals will display a frequency distribution, and the time of transition of the crop (the population as a whole) can be taken as the mean or median time of transition of the individual plants. The problem is primarily one of defining the interface between the two growth phases. The reproductive growth phase can be defined as the period of plant growth when a measurable proportion of the newly acquired plant dry matter is partitioned to the sexual reproductive organ, or its derivative. A botanically determined plant such as barley becomes committed to reproductive growth when its terminal, apical meristem starts to produce floret primordia. However, leaf-extension growth continues for some time after flower initiation and does not cease until, or shortly before, anthesis.

In the present context, it may be better to define the cessation of vegetative growth and commencement of reproductive growth for such a crop as the median time of anthesis. For crops such as soybean or mungbean, the time of transition may be usefully taken as the median flowering day (the day by which one-half of the population has flowered). There is no definitive time of transition from vegetative to reproductive growth. The experimentalist must have the flexibility to make his own judgement, a judgement which is necessarily tempered by his ability to be able to recognize and record the time of transition with the experimental resources available to him.

The inference of gross crop-growth rates

We can only measure directly the *net* amounts of crop dry matter present at any time. The *net* amount of dry matter is the difference between the *gross* amount produced and the amount lost through senescence, disease or pests. We cannot unambiguously attribute differences in the *net* growth rate of a crop to one or other of the five physiological determinants of growth unless we know the rate of loss of crop dry matter. To tackle this problem, we need to develop techniques for inferring the *gross* amounts, or rates, of dry matter produced. It is a problem that is fundamental to our establishing a proper understanding of the physiological limitations to crop growth.

In general, the most labile plant tissues are the leaves. Whereas healthy growing plants naturally shed their oldest leaves as new ones develop, they do not normally lose stems. Mechanical and pathological damage might cause stem loss, and many perennial and annual plants lose the stems bearing seed heads at, or after, the time that seeds reach maturity. However, as a useful first approximation we can assume that the standing stem dry weight throughout the vegetative growth phase represents the *gross* amount of stem material produced by the crop.

After cotyledon expansion, and before there is any visible leaf necrosis or loss, we can estimate the ratio of leaf to stem partition coefficients, η_L/η_S, from the ratio of the increments in leaf and stem dry weight. That is, we can estimate η_L/η_S as

$$\eta_L/\eta_S = \Delta W_L/\Delta W_S \tag{6.30}$$

where ΔW_L and ΔW_S represent the increments in leaf and stem dry weights during this period. At any later time, when there is visible leaf necrosis or loss, we can calculate the gross increment in leaf dry weight from the previously calculated ratio η_L/η_S and the measured increment in standing stem dry weight. The *gross* increment in above-ground dry weight, $(\Delta W_T)_{\text{gross}}$, can also be calculated as

$$(\Delta W_T)_{\text{gross}} = [1 + \eta_L/\eta_S] \Delta W_S. \tag{6.31}$$

Implicit in eqn (6.31) is the assumption that there are no ontogenetic or environmentally induced changes in the ratio η_L/η_S during the period of interest.

Ontogenetic changes in the ratio η_L/η_S are the most difficult to examine. If the crop being studied consists of plants, such as sorghum, which produce a succession of well-defined nodes, we can estimate the ratio η_L/η_S from measurements of the leaf to stem dry-weight ratios of successive nodes. We must be aware of any changes in the cropping environment during the production of the different nodes, and discount those nodes which display marked tissue necrosis. We can also examine environmentally induced changes in the ratio η_L/η_S by using eqn (6.30) to examine serially sown crops or plants.

The inference of gross crop-growth rates during reproductive growth is more difficult. With an indeterminate plant, such as tomato or melon, we can use the techniques described above by using the dry-weight ratios of leaf to stem and fruit to stem for each of the identifiable reproductive units ("nodes") on the plant. For a determined crop, such as barley, we can discount production of leaf dry matter during reproductive growth and estimate the gross increment in above-ground dry weight directly as the increment in standing stem and reproductive parts dry weight. However, some botanically indeterminate crop plants, such as kenaf, behave in an agriculturally determined way, that is they cease leaf production on flowering. Other botanically determinate plants, such as many cereal cultivars, behave in an agriculturally indeterminate way, by tiller production. Yet other crop plants, for example some soybean cultivars, behave in a mixed determinate/indeterminate way. They produce new leaf and stem extension growth during a part of their reproductive phase, becoming more determinate as the reproductive growth phase continues. It may be that for these crops we need to separate and distinguish the vegetative and reproductive meristems. The decision must remain the prerogative of the experimenter.

The collection of leaf litter from the ground beneath a crop provides only a *minimum* estimate of leaf dry-matter loss. It may underestimate the actual losses because of de-coupled, respiratory losses of dry matter by the shed tissues, microbial degradation of them or mechanical degradation by scavanging insects, etc.

The technique that is used to infer the *gross* amount, or rate, of dry-matter production by a crop will differ with crops and between the different growth phases of a particular crop. The technique used must depend upon the judgement of the experimenter. What needs to be emphasized is that estimates of *gross* dry-matter production are essential if the physiological limitations to crop growth are to be unambiguously identified and understood.

The determinants of crop growth

When we have estimated the *gross* amount, or rate, of dry-matter production during a growth phase, we can use eqns (6.1) and (6.2) to examine effectors of the growth determinants ε, J, η, V and t. The time period over which we make an estimate of any one of the determinants depends upon the experimental objectives. For example, if the physical environment of the crop changes rapidly and markedly during a particular growth phase and we wish to study the effect of these changes on one or more of the physiological growth determinants, we will need to make frequent harvests of crops during the growth phase. However, if we want to study the effects of different crop-management strategies on the determinants, fewer harvests, at the cardinal times during growth, may be adequate. From these we can estimate average values of the determinants during the growth phases. At the outset we need to define our experimental objectives with some precision.

An example of the application of the analysis to crop data is useful here. A soybean crop was harvested soon after emergence, at the start and end of flowering and at maturity. Cummulated incident radiation was recorded between harvests, and visual assessment of the proportion of the light intercepted by the crop at each harvest was made. The harvest data are given in Table 6.2. Between emergence and the start of flowering there was no visible

Table 6.2

Experimental harvest data for leaf (W_L), stem (W_S) and pod (W_P) dry weights (in g m^{-2}) at emergence, at the beginning and end of flowering, and at maturity of a soybean crop. The cummulated incident light (ΣS) (in MJ m^{-2}) during vegetative growth, flowering and maturity.

	W_L	W_S	W_P	ΣS
Emergence	4	3	—	—
First flowering	95	99	—	233
End of flowering	119	184	48	209
Maturity	14	162	317	336

leaf necrosis, and the ratio η_L/η_S was calculated, using eqn (6.30), as

$$\eta_L/\eta_S = \Delta W_L/\Delta W_S = 91/96 = 0.95. \tag{6.32}$$

Using this value of η_L/η_S, $(\Delta W_T)_{\text{gross}}$ (the gross increment in above-ground dry matter) was calculated during flowering using eqn (6.31), with an additional increment to account for pod and flower dry weight. Between the end of flowering and maturity it was assumed that no new stem or leaf tissue was produced. Estimated increments in the *gross* above-ground dry matter of the crop, the amount of light energy intercepted by the crop and the calculated efficiency of production of new above-ground dry matter, ε_T, are given in Table 6.3. This type of calculation can be used to establish whether or not differences in crop yield resulting from different management strategies, for example different plant densities, are a consequence of induced physiological differences in dry-matter partitioning or light-use efficiency, or are a consequence of physical effects such as the amount of the incident light energy intercepted by the crop.

Table 6.3

Estimated gross above-ground dry-matter increments ($(W_T)_{\text{gross}}$), amount of intercepted light energy ($\bar{Q}\Sigma S$) (in MJ m^{-2}) and efficiency of above-ground dry-matter production (ε_T) (in µg J^{-1}) during vegetative growth, flowering and maturity. (Data provided by Dr R. Lawn).

	$(W_T)_{\text{gross}}$	$\bar{Q}\Sigma S$	ε_T
Vegetative growth	187	155	1.2
Flowering	177	159	1.1
Maturity	269	229	1.2

Integrating studies on single plants

It is often argued that measurements made on single space plants, grown in controlled-environment facilities or under natural daylight, are not particularly relevant to an understanding of crop growth and performance in the field. The problem is primarily one of integrating information on the growth of spaced plants into an understanding of crop behaviour. Whereas some of the physiological determinants of crop growth are properties of the crop as a whole, for example the amount of light intercepted and the light-utilization efficiency, others are primarily properties of the component plants themselves, for example the time to flowering. However, even those determinants that are essentially properties of the crop as a whole contain a component attributable directly to the individual component plants.

We can write the instantaneous proportion of the incident light flux density intercepted by the crop, Q, as

$$Q = 1 - \exp(-kL) \tag{3.18}$$

where k is a canopy extinction coefficient and L is the leaf area index of the crop. The canopy light extinction, k, describes the architecture of the crop; it is a property of the crop as a whole. The leaf area index, L, can be re-written as

$$L = s_A W_L \tag{6.33}$$

where W_L is the standing leaf dry weight per unit ground area and s_A is the specific leaf area. The specific leaf area is a property of the component plants. It is also a function of the environment in which new leaves develop, and this environment may depend upon the crop structure and shading of new leaves developing within the leaf canopy. However, studies of environmentally induced changes in the specific leaf area of new, fully expanded leaves on spaced plants can provide useful information on, and understanding of, the effects of the aerial and rooting environments on the development of the leaf canopy. The effects of changes in the average daily light integral and temperature on s_A have been described in Chapter 3 (eqn 3.28).

It may be simpler, and less demanding in time and resources, to study ontogenetic and environmentally induced changes in the partition of dry matter on spaced single plants. For example, the analysis of the effects of water stress on the growth of the legume *M. atropurpureum* described in Section 2.1 was facilitated by the use of single spaced plants. The conclusions drawn from this study on the relative effects of water on the three growth determinants ε, η and V can be readily extrapolated to the field. Since water stress increased V (the rate of leaf-weight loss) in spaced plants, we may conclude that in the field such stress will reduce the maximum standing leaf dry weight of a forage crop of *M. atropurpureum*. This hypothesis is amenable to test *in the field*. If it is found to be correct, we can have some confidence in applying the understanding obtained for spaced plants to the field growth of the crop.

6.5   Conclusions

Synopsis: The logical analysis of the rate and amount of crop growth identifies five physiological determinants of growth which can be studied at two different levels—the agricultural or field level and the basic plant physiological research level. It provided a means of integrating physiological knowledge of plant processes into an understanding of the performance of agricultural crops.

The analyses described in this book do not pretend to constitute a definitive analysis of crop growth data. I have attempted to show how we might analyse data for field-grown crops into a number of multiplicative components (growth determinants) which are more amenable to analysis in terms of fundamental plant physiological processes. The five physiological determinants of growth are defined at one level in the biological hierarchy of organization of the crop. They are defined at the highest level, that of the whole functioning crop. By distinguishing several physiological determinants of growth at that level, the determinants themselves appear to be separately amenable to examination and analysis at lower levels in the biological hierarchy.

Perhaps it does not matter too much whether we take an evolutionary (Kuhn, 1963) or revolutionary (Popper, 1968) view of scientific progress. What is important is our perception of the hierarchical organization of our science, and our perception of the problem that we are researching.

Passioura (1979) and Thornley (1980) have discussed the hierarchical organization of biological systems and examined the implications of their perceived hierarchy to research in the plant sciences. The scheme which they both discuss is based on an increasing material complexity of components at successively higher levels. This hierarchy, which we can describe as a state hierarchy, can be written as:

> plant community
> single plant
> organ (leaf, stem, fruit, root, etc.)
> tissue
> cell

and their discussions centre around the relevance of research at any one level to the development of an understanding of a phenomenon at another, higher level. The main thrusts of their arguments are:

> (i) discussions or descriptions at one level may be connected to the next higher level in an explanatory or mechanistic scheme;

(ii) the relationship between adjacent levels is not symmetric, and whereas the successful operation of the higher of two levels requires the lower level to function effectively, the converse is not true.

In his essay "The architecture of complexity", Simon (1962, p. 479) wrote:

> We pose a problem by giving the state description of the solution. The task is to discover a sequence of processes that will produce the goal state from an initial state. Translation from the process description to the state description enables us to recognize when we have succeeded . . . The general paradigm is: given a blueprint, to find the corresponding recipe. Much of the activity of science is an application of that paradigm: given the description of some natural phenomena, to find the differential equations for processes that will produce the phenomena.

Let us radically redefine the hierarchical organization of plant systems. Let us not define it by a series of material states, but by the degree of the constraint operating between adjacent material states. We can do so by defining a process hierarchy, which we can write as:

biological process
physiological process
chemical, physical or biochemical process
molecular process

We need to define what is meant by the terms "biological process" and "physiological process". We can define the biological process as growth, that is an increase in the mass or numbers of an individual. We could extend this definition to include the capacity for growth, the capacity of an organism to adapt or evolve in order to enable it to grow and survive in as wide a range of environments as possible. A physiological process is an expression of a matrix of chemical and physical processes. It is confined, both in time and space, within a biological structure, and could not take place outside of that structure. It is worthwhile quoting Simon (p. 481) again:

> The notion of substituting a process description for a state description of nature has played a central role in the development of modern science. Dynamic laws, expressed in the forms of systems of differential or difference equations, have in a large number of cases provided the clue for the simple description of the complex . . . The correlation between state description and process description is basic to the functioning of any adaptive organism, to its capacity for acting purposefully upon its environment.

Passioura (1979) has observed that most research in the plant sciences is undertaken at the highest and lowest levels in the material state hierarchy.

He implies that this asymmetry in the distribution of research effort is unhealthy. Perhaps it reflects an intuitive feeling on the part of many plant scientists that the recognition of a process hierarchy is more important than the recognition of a material state hierarchy to the development of an understanding about biological systems.

The physiological analysis of crop growth described in this book is a description of growth at one level in a process hierarchy. The five physiological determinants of crop growth represent five distinct physiological processes. The transition to a state description of the cropping system is not made until the final intergration (cf. eqn 1.3).

Because the many species of crop plants differ so widely in their physical structure and physiological behaviour, it is not practical to attempt to describe a minimum data set which will allow the analysis to be applied to all crops. The analysis represents a philosophical, conceptual framework within which we might start to develop an understanding of the factors, both environmental and physiological, which determine the rate and amount of crop growth. It has to be left to the judgement of the individual crop scientist to decide what information on his crop he might most profitably collect. That will depend upon the objectives of his experiment and the limitations of the resources avilable to him. I have sought to do no more than indicate how he might proceed if he chooses to conduct his research within the framework of this analysis. The analysis itself provides no answer to the problems, but it does provide a rational means of ordering knowledge and information. Understanding derives from the process of ordering knowledge.

Appendixes

A

Definitions

closed crop—a closed crop is a leaf canopy within which we can neglect spatial heterogeneity in the downward light flux density incident upon a horizontal plane at any horizon within the canopy. We can assume that variation in the downward light flux density is confined to the vertical plane. The downward light flux density decreases with increasing depth below the canopy's uppermost surface according to the Beer–Lambert Law. Beneath any cumulative leaf area index, L, the downward light flux density, I, is related to the light flux density incident upon the canopies uppermost surface, I_0, according to $I = I_0 \exp(-kL)$, where k is a canopy light extinction coefficient.

extensive cropping system —an extensive cropping system is defined as one for which there is marked spatial heterogeneity in either its physical environment or in its botanical composition. An arable cropping system covering large tracts of land, perhaps a large wheat crop, may be an extensive system. By way of contrast, a well-cultivated, inter-cropping system or a small pastoral system would also be extensive cropping systems.

heuristic model—a heuristic model is one which enables the modeller to discover things about the system he is modelling that are not immediately obvious to him as properties of the system's components. For example, the properties of positive and negative feedback emerge as properties

143

of systems comprised of two or more components, but are not obvious
when the performances of the components are examined in isolation.
A model of such a system could be built as a heuristic aid to help the
modeller see what the emergent properties of the system were.

intensive cropping system—an intensive cropping system is one for which
spatial heterogeneity in either the physical environment or botanical
composition of the system can be discounted, or explicitly allowed for,
in the analysis of the system.

knowledge—knowledge is the cognizance of a proven fact or observation.

leaf area index—the leaf area index of a crop is the ratio of the actual leaf
area of the crop to the land area covered by the crop, projected downward onto a horizontal surface beneath the crop.

light energy—light energy is defined throughout this book as that portion
of the radiant energy that lies within the 400–700 nm waveband, the
conventional definition of photosynthetically active radiation. The daily
light integral is the daily integral of the downward light flux density
incident upon an unobstructed horizontal surface at ground level. The
mean daily light flux density is the ratio of the daily light integral and
the daylength.

light-utilization efficiency—the light-utilization efficiency of a plant or a
crop is the ratio of the amount of *new* dry matter produced by the plant
or crop for each unit of light energy intercepted by it.

maximum potential yield—the maximum potential yield of a crop is defined
as the yield of the harvestable component of the crop in the absence of
water or mineral/nutrient deficiencies when losses of that component
through physiological, pathological and mechanical causes have been
minimized.

ontogeny—the ontogeny of a plant is the history of its development during
its entire life.

partition coefficient—the partition coefficient for an organ is the ratio of the
gross dry-matter increment of a particular organ to the total gross dry-
matter increment of the whole plant.

phenology—phenology is the study of periodical phenomena of plants. It
includes studies such as the study of flowering by the plant in relation
to the plant's environment.

procrustean—a procrustean assumption is one that simplifies by drastic
means. It is a means of "cutting the Gordian knot"; that is, a way of
getting out of difficulty by the shortest and most drastic means.

specific activity—the specific activity of an organ is its activity expressed
per unit mass of the organ.

specific leaf area—the specific leaf area is the leaf area subtended by unit
dry weight of leaf tissue.

tautology—a tautology is the repetition of the same idea in different words.

teleology—teleology is the doctrine or theory that all things or processes were designed to fulfil a purpose. It implies a sense of purpose and even cognizance. In an evolutionary context, teleological arguments seem to me to be quite satisfactory. In an evolutionary context, it is likely that if a plant function had no purpose it would have been selected out of the population, and would not exist now as a discrete and identifiable function.

understanding—understanding is an explanation of a phenomenon observed at one level in the biological hierarchy of organization of a crop or plant that is based on knowledge acquired at some lower level in the hierarchy of the organization.

B

Main Symbols

I have tried to use symbols in a consistent way throughout this book. This list represents the main symbols that have been used.

Independent variables

R	daily integral of the radiant energy flux density incident upon a horizontal surface at ground level	$MJ\,m^{-2}\,d^{-1}$
R_M	annual mean value of the daily integral of the radiant energy flux density	$MJ\,m^{-2}\,d^{-1}$
R_D	amplitude of the seasonal variation in the daily integral of the radiant energy flux density	$MJ\,m^{-2}\,d^{-1}$
S	daily integral of the light flux density incident upon a horizontal surface at ground level	$MJ\,m^{-2}\,d^{-1}$
$\bar{T}$	average daily air temperature	$°C$
$\bar{T}_M$	mean annual value of the average daily air temperature	$°C$
h	daylength	seconds
N	latitude	degrees (N or S)

N	nitrogen concentration	$g\,(N)\,m^{-3}$
I_0	unobstructed downward light flux density	$J\,m^{-2}\,s^{-1}$
$\bar{I}$	mean daily incident downward light flux density	$J\,m^{-2}\,s^{-1}$

Dependent variables

W	total plant or crop dry weight	g or $g\,m^{-2}$
W_H	dry weight of the harvestable component of a plant or crop	g or $g\,m^{-2}$
A	total plant leaf area	m^2
L	leaf area index	
V	daily rate of loss of dry matter	$g\,d^{-1}$ or $g\,m^{-2}\,d^{-1}$

The subscripts, T, L, S, R and F refer to above-ground parts, leaf, stem, root are reproductive parts of the plant or crop respectively.

Parameters and constants

Q	the instantaneous proportion of the light flux density incident upon a plant or a crop that is intercepted by it.	
$\bar{Q}$	the average proportion of the incident light flux density intercepted by a plant or crop over an extended period of time.	
J	the amount of light energy intercepted by a plant or crop during the course of the day	$MJ\,m^{-2}\,d^{-1}$
E	the cumulated amount of light energy intercepted by a plant or crop	$MJ\,m^{-2}$
μ	specific growth rate	$g\,g^{-1}\,d^{-1}$
μ^*	"modified" specific growth rate (growth rate per unit of leaf dry weight)	$g\,g^{-1}\,d^{-1}$
σ	specific activity of an organ	$g\,(\ )\,g\,(\ )^{-1}\,d^{-1}$
η	partition coefficient	
η_H	portion of new dry matter partitioned to the harvestable component of the crop	
ε	light-utilization efficiency	$\mu g\,J^{-1}$

s_A	specific leaf area	$m^2 g^{-1}$
F_A	leaf-area ratio	$m^2 g^{-1}$
F_{max}	rate of light-saturated leaf photosynthesis	$g m^{-2} s^{-1}$
∇_C	gross daily photosynthetic integral	$g (CO_2) m^{-2} d^{-1}$
∇_F	net daily photosynthetic integral	$g (CO_2) m^{-2} d^{-1}$
∇_R	daily respiratory integral	$g (CO_2) m^{-2} d^{-1}$
γ	abscission or dry-matter loss constant	g^{-1}
α	leaf photochemical efficiency	$\mu g (CO_2) J^{-1}$
k	canopy light extinction coefficient	
f_M	fraction of the element M in plant dry matter	$g (M) g (dry-matter)$
q	conversion factor to convert $g (CO_2)$ to g (dry matter)	
Y_g	growth yield constant	

The subscripts T, L, S and R refer to above-ground parts of the crop, leaves, stems and roots respectively.

The subscripts N, P and C refer to elemental compositions and specific organ activities with respect to nitrogen, phosphorus and carbon.

C

Some Conversion Factors

Light measurements

$$1 \text{ MJ} \equiv 278 \text{ watt hours} \equiv 2.39 \times 10^5 \text{ cals}$$

$$1 \text{ J m}^{-2}\text{s}^{-1} \equiv 1 \text{ W m}^{-2} \equiv 1.43 \times 10^{-3} \text{ cal min}^{-1} \text{ cm}^{-2}$$

1 Einstein of natural daylight is approximately equivalent to 0.23 MJ

Yield measurements

$$1 \text{ t ha}^{-1} \equiv 1000 \text{ kg ha}^{-1} \equiv 0.1 \text{ kg m}^{-2} \equiv 100 \text{ g m}^{-2}$$

Photosynthesis measurements

$$1 \times 10^{-3} \text{ g}(CO_2)\text{m}^{-2}\text{s}^{-1} \equiv 3.6 \times 10^{-2} \text{ g}(CO_2)\text{dm}^{-2}\text{hr}^{-1}$$
$$\equiv 2.27 \times 10^{-5} \text{ mole}(CO_2)\text{m}^{-2}\text{s}^{-1} \text{ (at STP)}$$

Gaseous concentrations

$$300 \text{ vpm gaseous } CO_2 \equiv 0.59 \text{ g}(CO_2)\text{m}^{-3} \text{ (at STP)}$$

$$1000 \text{ vpm gaseous } CO_2 \equiv 1.95 \text{ g}(CO_2)\text{m}^{-3} \text{ (at STP)}$$

$$1\% \text{ gaseous } O_2 \equiv 14.3 \text{ g}(O_2)\text{m}^{-3} \text{ (at STP)}$$

$$20\% \text{ gaseous } O_2 \equiv 286 \text{ g}(O_2)\text{m}^{-3} \text{ (at STP)}$$

References

Acock, B., Hand, D. W., Thornley, J. H. M., and Warren Wilson, J. (1976). Photosynthesis in stands of green peppers: an application of empirical and mechanistic models to controlled environment data. *Ann. Bot.* **40**, 1293–1307.

Acock, B., Charles-Edwards, D. A., Fitter, D. J., Hand, D. W., Ludwig, L. J., Warren Wilson, J., and Withers, A. C. (1978a). The contribution of leaves from different levels within a tomato crop to canopy net photosynthesis: an experimental examination of two canopy models. *J. Exp. Bot.* **29**, 815–827.

Acock, B., Charles-Edwards, D. A., Fitter, D. J., Hand, D. W., and Warren Wilson, J. (1978b). Modelling canopy net photosynthesis by isolated blocks and rows of chrysanthemum plants. *Ann. Appl. Biol.* **90**, 255–263.

Acock, B., Charles-Edwards, D. A., and Sawyer, S. (1979). Growth response of a chrysanthemum crop to environment. III. Effects of radiation and temperature on dry-matter partitioning and photosynthesis. *Ann. Bot.* **44**, 289–300.

Allen, E. J., and Scott, R. K. (1980). An analysis of the growth of the potato crop. *J. Agric. Sci., Camb.* **94**, 583–606.

Antonovics, J. (1968). The population genetics of mixtures. *In* "Plant Relations in Pastures" (J. Wilson, ed.), pp. 233–252. CSIRO, Melbourne.

Austin, R. B., Bingham, J., Blackwell, R. D., Evans, L. T., Ford, M. A., Morgan, C. L., and Taylor, M. (1980). Genetic improvements in winter wheat yields since 1900 and associated physiological changes. *J. Agric. Sci., Camb.*, **94**, 675–689.

Barnes, A. (1977). The influence of the length of the growth period and planting density on total crop yield. *Ann. Bot.* **41**, 883–895.

153

Barnes, A., and Hole, C. C. (1978). A theoretical basis of growth and maintenance respiration. *Ann. Bot.* **42**, 1217–1221.

Biscoe, P. V., and Gallagher, J. N. (1977). Weather, dry-matter production and yield. *In* "Environmental Effects on Crop Physiology" (J. J. Landsberg and C. V. Cutting, eds.), pp. 75–100. Academic Press, London.

Chanter, D. O. C. (1981). The use and misuse of linear regression methods in crop modelling. *In* "Mathematics and Plant Physiology" (D. A. Rose and D. A. Charles-Edwards, eds.), pp. 252–267. Academic Press, London.

Charles-Edwards, D. A. (1976). Shoot and root activities during steady-state plant growth. *Ann. Bot.* **40**, 767–772.

Charles-Edwards, D. A. (1978). An analysis of the photosynthesis and productivity of vegetative crops in the U.K. *Ann. Bot.* **42**, 717–731.

Charles-Edwards, D. A. (1979). Photosynthesis and crop growth. *In* "Photosynthesis and Plant Development (R. Marcelle, H. Clijsters and M. van Pouke, eds.), pp. 111–124. Junk, The Hague.

Charles-Edwards, D. A. (1981). "The Mathematics of Photosynthesis and Productivity." Academic Press, London.

Charles-Edwards, D. A., and Acock, B. (1977). Growth response of a chrysanthemum crop to the environment. II. A mathematical analysis relating photosynthesis and growth. *Ann. Bot.* **41**, 49–58.

Charles-Edwards, D. A., and Thorpe, M. R. (1976). Interception of diffuse and direct beam radiation by a hedgerous apple orchard. *Ann. Bot.* **40**, 603–613.

Charles-Edwards, D. A., Charles-Edwards, J., and Sant, F. I. (1974). Leaf photosynthetic activity in six temperate grass varieties grown in contrasting light and temperature environments. *J. Expt. Bot.* **25**, 715–724.

Cooper, J. P. (1970). Potential production and energy conversion in temperate and tropical grasses. *Herbage Abstracts* **40**, 1–15.

Criswell, J. G., and Shibles, R. M. (1971). Physiological basis for genotypic variation in net photosynthesis of oat leaves. *Crop Sci.* **11**, 550–553.

Davidson, R. L. (1969). Effect of root/leaf temperature on root/shoot ratios in some pasture grasses and clover. *Ann. Bot.* **33**, 561–569.

de Wit, C. T. (1960). On competition. *Versl. Landbouwkd. Onderz.* **16**, 1–82.

Donald, C. M., and Hamblin, J. (1976). Biological yield and harvest index of cereals as agronomic and plant breeding criteria. *Adv. Agron.* **28**, 361–405.

Dornhoff, G. M., and Shibles, R. M. (1970). Varietal differences in net photosynthesis in soyabean leaves. *Crop Sci.* **10**, 42–45.

Dornhoff, G. M., and Shibles, R. M. (1976). Leaf morphology and anatomy in relation to CO_2 exchange rate of soybean leaves. *Crop Sci.* **16**, 377–381.

Enyi, B. A. C. (1973). Effect of population on growth and yield of soya bean (*Glycine max*). *J. Agric. Sci., Camb.* **81**, 131–138.

Fisher, M. J., and Charles-Edwards, D. A. (1982). A physiological approach to the analysis of crop growth data. III. The effects of repeated short-term soil water deficits on the growth of spaced plants of the legume *Macroptilium atropurpureum* cv. Siratro. *Ann. Bot.* **49**, 341–346.

Fisher, M. J., Charles-Edwards, D. A., and Campbell, W. A. (1980). A physiological approach to the analysis of crop growth data. II. Growth of *Stylosanthes humilis*. *Ann. Bot.* **46**, 425–434.

Fisher, M. J., Charles-Edwards, D. A., and Ludlow, M. M. (1981). An analysis of the effects of repeated short-term soil water deficits on stomatal conductance to carbon dioxide and leaf photosynthesis by the legume *Macroptilium atropurpureum* cv. Siratro. *Aust. J. Plant Physiol.* **8**, 347–357.

Gardener, C. J., and Rathjen, A. J. (1975). The different response of barley genotypes to nitrogen application in a Mediterranean type climate. *Aust. J. Agric. Res.* **26**, 219–230.

Greenwood, D. J. (1981). Crop response to agricultural practice. *In* "Mathematics and Plant Physiology (D. A. Rose and D. A. Charles-Edwards, eds.), pp. 195–216. Academic Press, London.

Greenwood, D. J., and Barnes, A. (1978). A theoretical model for the decline in the protein content in plants during growth. *J. Agric. Sci., Camb.* **91**, 461–466.

Greenwood, D. J., Cleaver, T. J., and Turner, M. K. (1974). Fertilizer requirements of vegetable crops. The Fertilizer Society, Proceedings No. 145, Alembic House, 93 Albert Embankment, London SE1 7TU.

Greenwood, D. J., Cleaver, T. J., Loquens, S. M. H., and Niendorf, K. B. (1977). Relationship between plant weight and growing period for vegetable crops in the U.K. *Ann. Bot.* **41**, 987–997.

Greenwood, D. J., Barnes, A., and Leaver, T. J. (1978). Measurement and prediction of the changes in protein contents of field crops during growth. *J. Agric. Sci., Camb.* **91**, 467–477.

Gulman, S. L., and Chu, C. C. (1981). The effects of light and nitrogen on photosynthesis, leaf characteristics and dry-matter accumulation in the chaparal shrub *Diplacus auranticas. Oecologia (Berl.),* **49**, 207–212.

Hall, R. L. (1974). Analysis of the nature of interference between plants of different species. I. Concepts and extension of the de Wit analysis to examine effects. *Aust. J. Agric. Res.* **25**, 739–747.

Hansen, G. K., and Jensen, C. R. (1977). Growth and maintenance respiration in whole plants, tops and roots of *Lolium multiflorum. Physiol. Plant.* **39**, 155–164.

Harper, F., and Compton, I. J. (1980). Sowing date, harvest date and the yield of forage brassica crops. *Grass and Forage Science* **35**, 147–157.

Heichel, G. M., and Musgrave, R. B. (1969). Relation of CO_2 compensation concentration to apparent photosynthesis in maize. *Plant Physiol.* **44**, 1724–1728.

Hunt, R. (1975). Further observations on root–shoot equilibria in perennial ryegrass (*Lolium perenne L.*). *Ann. Bot.* **39**, 745–755.

Hunt, W. F., and Loomis, R. S. (1979). Respiration modelling and hypothesis testing with a dynamic model of sugar beet growth. *Ann. Bot.* **44**, 5–17.

Irvine, J. E. (1967). Photosynthesis in sugar cane varieties under field conditions. *Crop Sci.* **7**, 297–300.

Imai, K., and Murata, Y. (1979a). Effect of carbon dioxide concentration on growth and dry-matter production of crop plants. VI. Effect of oxygen concentration on the carbon dioxide/dry-matter production relationship in some C_3 and C_4 crop species. *Japan. J. Crop Sci.* **48**, 58–65.

Imai, K., and Murata, Y. (1979b). Effect of carbon dioxide concentration on growth and dry-matter production of crop plants. VII. Influence of light intensity and temperature on the effect of carbon dioxide enrichment in some C_3 and C_4 species. *Japan. J. Crop Sci.* **48**, 409–417.

Jackson, J. E., and Palmer, J. W. (1979). A simple model of light transmission and interception by discontinuous canopies. *Ann. Bot.* **44**, 381–383.

Jackson, J. E., and Palmer, J. W. (1981). Light distribution in discontinuous canopies: calculation of leaf areas, canopy volumes above defined 'irradiance contours' for use in productivity modelling. *Ann. Bot.* **47**, 561–565.

Jones, P. G., and Laing, D. R. (1978). The effects of phenological and meteorological factors on soybean yield. *Agr. Meteorol.* **19**, 485–495.

Kasanaga, H., and Monsi, M. (1954). On the light-transmission of leaves and its meaning for the production of dry-matter in plant communities. *Jap. J. Bot.* **14**, 304–324.

Kuhn, T. S. (1963). "The Structure of Scientific Revolutions." Chicago University Press, Chicago.

Loomis, R. S., and Williams, W. A. (1963). Maximum crop productivity: an estimate. *Crop Sci.* **3**, 67–72.

Luckwill, L. C. (1960). The physiological relationship of root and shoot. *Scient. Hort.* **14**, 22–26.

Ludlow, M. M., and Charles-Edwards, D. A. (1980). Analysis of the regrowth of a tropical grass/legume sward subjected to different frequencies and intensities of defoliation. *Aust. J. Agric. Res.* **31**, 673–692.

Ludlow, M. M., Stobbs, T. H., Davis, R., and Charles-Edwards, D. A. (1982). Effect of sward structure in two tropical grasses with contrasting canopies on light distribution, net photosynthesis and the size of bite harvested by grazing cattle. *Aust. J. Agric. Res.* **33**, 187–201.

McCown, R. L. (1981a). The climatic potential for beef cattle production in tropical Australia. I. Simulating the annual cycle of liveweight change. *Agric. Systems* **6**, 303–317.

McCown, R. L. (1981b). The climatic potential for beef cattle production in tropical Australia. III. Variation in the commencement, cessation and duration of the green season. *Agric. Systems* **7**, 163–178.

McCree, K. J. (1970). An equation for the rate of respiration of white clover plants grown under controlled conditions. *In* "Prediction and Measurement of Photosynthetic Productivity." (Proc. IBP/PP Technical Meeting, Trebon), Centre for Agricultural Publishing, Wageningen.

McGilchrist, C. A., and Trenbath, C. R. (1971). A revised analysis of plant competition experiments. *Biometrics* **27**, 659–671.

Monteith, J. L. (1965). Light distribution and photosynthesis in field crops. *Ann. Bot.* **29**, 17–37.

Monteith, J. L. (1977). Climate and efficiency of crop production in Britain. *Phil. Trans. R. Soc. Lond. B.* **281**, 277–294.

Mooney, K. A., Ferrar, P. J., and Slatyer, R. O. (1978). Photosynthetic capacity and carbon allocation patterns in diverse growth forms of *Eucalyptus*. *Oecologia (Berl.)* **36**, 103–111.

Muchow, R. C., and Charles-Edwards, D. A. (1982a). A physiological analysis of growth of mungbeans at a range of plant densities in tropical Australia. I. Dry matter production. *Aust. J. Agric. Res.* **33**, 41–51.

Muchow, R. C., and Charles-Edwards, D. A. (1982b). A physiological analysis of growth of mungbeans at a range of plant densities in tropical Australia. II. Seed production. *Aust. J. Agric. Res.* **33**, 53–61.

Nass, H. G. (1973). Determination of characters for yield selection in spring wheat. *Can. J. Plant Sci.* **53**, 755–762.

Natarajan, M., and Willey, R. W. (1980a). Sorghum–pigeon pea intercropping and the effects of plant population density. I. Growth and yield. *J. Agric. Sci., Camb.* **95**, 51–58.

Natarajan, M., and Willey, R. W. (1980b). Sorghum–pigeon pea intercropping and the effects of plant population density. II. Resource use. *J. Agric. Sci., Camb.* **95**, 59–65.

Nix, N. A., and Fitzpatrick, E. A. (1969). An index of crop water stress related to wheat and grain sorghum yields. *Agr. Meteorol.* **6**, 321–337.

Passioura, J. B. (1973). Sense and nonsense in crop simulation. *J. Aust. Inst. Agric. Sci.* **39**, 181–183.

Passioura, J. B. (1979). Accountability, philosophy and plant physiology. *Search* **10**, 347–350.

Pearce, R. B., Charlson, C. E., Barnes, D. K., Hart, R. H., and Hanson, C. H. (1969). Specific leaf weight and photosynthesis in alfalfa. *Crop Sci.* **9**, 423–426.

Popper, K. R. (1968). "The Logic of Scientific Discovery." Hutchinson, London.

Potter, J. R., and Jones, J. W. (1977). Leaf area partitioning as an important factor in growth. *Plant Physiol.* **59**, 10–14.

Prioul, J., and Bourdu, R. (1973). Graphical display of photosynthetic adaptability to irradiance. *Photosynthetica* **7**, 405–407.

Riggs, D. S. (1970). "The Mathematical Approach to Physiological Problems." MIT Press, Massachusetts.

Riggs, T. J., Hanson, P. R., Start, N. D., Miles, D. M., Morgan, C. L., and Ford, M. A. (1981). Comparison of spring barley varieties grown in England and Wales between 1880 and 1890. *J. Agric. Sci., Camb.* **97**, 599–610.

Saeki, T. (1960). Interrelationships between leaf amount, light distribution and total photosynthesis in a plant community. *Bot. Mag., Tokyo* **73**, 55.

Sibma, L. (1968). Growth of closed green crop surfaces in the Netherlands. *Neth. J. Agric. Sci.* **16**, 211–216.

Simon, H. A. (1962). The architecture of complexity. *Proc. Amer. Philosophical Soc.* **106**, 467–482.

Singh, I. D., and Stoskopf, N. C. (1971). Harvest index in cereals. *Agron. J.* **63**, 224–226.

Tateno, K. and Iida, K. (1980). Seasonal variations of leaf photosynthesis in Italian ryegrass and orchardgrass. *J. Japan. Grassl. Soc.* **26**, 297–304.

Thornley, J. H. M. (1976). "Mathematical Models in Plant Physiology." Academic Press, London.

Thornley, J. H. M. (1977). Growth, maintenance and respiration: a re-interpretation. *Ann. Bot.* **41**, 1191–1203.

Thornley, J. H. M. (1977). Root: shoot interactions. *In* "Integration of Activity in the Higher Plants (D. M. Jennings, ed.), Cambridge University Press.

Thornley, J. H. M. (1978). Crop response to fertilizer. *Ann. Bot.* **42**, 817–826.

Thornley, J. H. M. (1980). Research strategy in the plant sciences. *Plant, Cell and Environment* **3**, 233–236.

Trenbath, B. R. (1968). Models and the interpretation of mixture experiments. *In* "Plant Relations in Pastures." (J. Wilson, ed.), pp. 145–162. CSIRO, Melbourne.

Troughton, A. (1960). Further studies on the relationship between shoot and root systems of grasses. *J. Br. Grassland Soc.* **15**, 41–47.

Vong, N. Q., and Murata, Y. (1978). Studies on the physiological characteristics of C_3 and C_4 crop species. II. The effects of air temperature and solar radiation on the dry-matter production of some crops. *Japan. J. Crop Sci.* **47**, 90–100.

Warren Wilson, J. (1971). Maximum yield potential. *In* "Transition from Extensive to Intensive Agriculture with Fertilizers." Proc. 7th Colloquim Int. Potash Inst., IPI, Berne.

Watson, D. J., Thone, G. V., and French, S. A. W. (1963). Analysis of growth and yield of winter and spring wheats. *Ann. Bot.* **27**, 1–22.

Index

159